BASIC ANALYSIS OF VARIANCE PROGRAMS FOR MICROCOMPUTERS

BASIC ANALYSIS OF VARIANCE PROGRAMS FOR MICROCOMPUTERS

Allen H. Wolach, *Illinois Institute of Technology*

Brooks/Cole Publishing Company
Monterey, California

This book is affectionately dedicated to

my parents

Brooks/Cole Publishing Company
A Division of Wadsworth, Inc.

Printed in the United States of America

10 9 8 7 6 5 4 3 2 1

Library of Congress Cataloging in Publication Data

Wolach, Allen H.
Basic analysis of variance programs for microcomputers.

Bibliography: p.
Includes index.
1. Analysis of variance--Computer programs.
2. Microcomputers--Programming. 3. Basic (Computer program language) I. Title.
QA279.W64 1982 519.5'352'0285425 82-14705

ISBN 0-534-01423-2

Subject Editor: C. Deborah Laughton
Production Coordinator: Louise Rixey
Cover Design: Vernon Boes

Preface

This text provides a series of analysis of variance programs and related programs (e.g., Latin square, analysis of covariance) that can be run on virtually all small microcomputer systems. The programs use a limited subset of the BASIC language. The subset of BASIC was chosen so that almost every microcomputer (Radio Shack, Apple, etc.) can run the programs with few modifications. Users simply select the appropriate program and enter it into the microcomputer--statement by statement. The program can then be recorded for future use.

Chapter 1 provides information about microcomputer systems. Some microcomputers have a few unique commands. In addition, some microcomputers allow as few as 40 characters on a line of the terminal screen. One can easily learn to get around the idiosyncrasies of a given microcomputer.

Investigators often know what analysis of variance they want, but cannot recognize the analysis of variance when it is presented with an alternate name. Chapter 2 presents most of the common alternate names for each of the analyses in this text. Chapter 2 also provides a list of text books that explain each of the analysis of variance procedures. The pages that cover a given analysis of variance are listed for each text book. The text books were selected because they are representative of approaches used for explaining analyses of variance. Although the list of text books may seem weighted toward the social and behavioral sciences, these text books are often used by engineers, physicists, and chemists.

Chapter 3 provides an example of data entry for each analysis of variance. A sample of the output from each program is also presented. The examples were carefully selected to insure that none of the factors in a given analysis of variance has the same number of levels as other factors. This makes it easy to learn the order in which data are entered. Directions for using a program are more complete than the directions for most statistical programs, including programs in standard "statistical packages."

All of the programs calculate the means for main effects and interactions. Thus, one can use these means to quickly perform follow-up tests (simple main effects tests, and procedures such as Tukey's procedure). That is, the program provides the means and error terms (from the analysis of variance) for these follow-up tests.

The analysis of variance programs were written with all factors other than blocks (subjects) factors as fixed effects factors. Most analyses of variance reported in journals use the fixed effects model. If an analysis of variance uses a random model for some or all of the factors, the mean square column of the analysis of variance summary table provides the necessary information for the denominators of the F-ratios.

The analysis of variance programs and related programs are presented in Chapter 4. Since completely randomized and completely randomized factorial analyses of variance can be performed with unequal numbers of subjects per group or subgroup, the first three programs are for a completely randomized analysis of variance, CR-K (Program 1), completely randomized factorial analysis of variance--two factor, CRF-JK (Program 2), and a completely randomized factorial analysis of variance--three factor, CRF-JKL (Program 3).[1] The remaining analysis of variance programs require an equal n for each cell. Program 4 calculates one factor analyses of variance--completely randomized analysis of variance, CR-K, or randomized block, RB-K. Program 5 calculates two factor analyses of variance--completely randomized factorial analysis of variance, CRF-JK, split plot analysis of variance, SPF-J.K, or randomized block factorial analysis of variance, RBF-.JK. Program 6 calculates three factor analyses of variance--completely randomized factorial analysis of variance, CRF-JKL, split plot analysis of variance, SPF-JK.L, split plot

[1]Note that J, K, and L refer to the levels of Factors A, B, and C, respectively. Some authors use P, Q, and R to represent levels of Factors A, B, and C.

analysis of variance, SPF-J.KL, or randomized block factorial analysis of variance, RBF-.JKL.

Some microcomputer systems may not have enough memory to hold multipurpose programs such as programs 4, 5, and 6. These programs have been separated into programs that only perform one analysis of variance. These analyses of variance are presented in Programs 7 through 15. Program 16 is for a Latin square analysis of variance (one variable of interest and two nuisance variables). Finally, Program 17 is for an analysis of covariance with one factor. The 17 programs can be summarized as follows:

Program 1. Completely randomized analysis of variance (does not require equal n in all cells), CR-K

Program 2. Completely randomized factorial analysis of variance (does not require equal n in all cells), CRF-JK

Program 3. Completely randomized factorial analysis of variance (does not require equal n in all cells), CRF-JKL

Program 4. A combined program that performs the analyses described for Programs 7 and 8

Program 5. A combined program that performs the analyses described for Programs 9, 10, and 11

Program 6. A combined program that performs the analyses described for Programs 12, 13, 14, and 15

Program 7. Completely randomized analysis of variance (requires equal n in all cells), CR-K

Program 8. Randomized block analysis of variance (requires equal in in all cells), RB-K

Program 9. Completely randomized factorial analysis of variance (requires equal n in all cells), CRF-JK

Program 10. Split-plot analysis of variance (requires equal n in all cells), SPF-J.K

Program 11. Randomized block factorial analysis of variance (requires equal n in all cells), RBF-.JK

Program 12. Completely randomized factorial analysis of variance (requires equal n in all cells), CRF-JKL

Program 13. Split-plot analysis of variance (requires equal n in all cells), SPF-JK.L

Program 14. Split-plot analysis of variance (requires equal n in all cells), SPF-J.KL

Program 15. Randomized block factorial analysis of variance (requires equal n in all cells), RBF-.JKL

Program 16. Latin square design (requires equal n in all cells), LS-K

Program 17. Analysis of covariance--one factor (requires equal n in all cells)

If the reader has any difficulty identifying an analysis by the name or symbol (e.g., CRF-JK) presented above, Chapter 2 presents the common alternate names for each analysis.

Programs 7 through 15 were written to use the same kind of data entry as the analyses of variance in programs 4 through 6. For example, one would enter the data in the same way for a split-plot analysis of variance (SPF-JK.L) in Program 6 or Program 13. Whenever possible the statement numbers in Programs 4 through 6 are identical to corresponding statement numbers in Programs 7 through 15. This makes it possible to observe how analysis of variance programs can be combined into a larger program.

All of the programs use difference formulas as opposed to computational formulas. Microcomputers that are programmed in BASIC make calculations to a limited number of digits (usually nine). Computational formulas produce intermediate results that are often out of range of the nine digit maximum for most versions of microcomputer BASIC. Chapter 5 shows the difference formulas that are used for the analysis of variance programs. No intermediate calculation for a sum of squares in the present programs is larger than the sum of squares total for the analysis of variance in question.

Most analysis of variance programs store data in arrays. In fact, they use arrays to store the data for intermediate tables that summarize each interaction. Arrays have two disadvantages. First, some microcomputer systems have limited array handling capabilities. For example, some versions of BASIC cannot handle arrays with more than two dimensions. Secondly, arrays take a great deal of computer memory. The present programs use no arrays. Observations are stored in DATA statements. Observations are reaccessed by the program through the use of the RESTORE statement. Since arrays are avoided, most of the programs can run relatively large data sets with less than 8k of random access memory. The number of statements for a program is increased because arrays are not used. However, one more than makes up for the increase in statements by not having to reserve memory for arrays.

The author would like to thank Dr. Maureen A. McHale of Northwestern State University of Louisiana, and Dr. Sylvia Van Berschot of Illinois Institute of Technology for their thoughtful reviews of the manuscript. My brother, Marshall W. Wolach, also made useful suggestions.

C. Deborah Laughton of Brooks/Cole did a superb job as editor for this book. Her suggestions have greatly improved the manuscript. I would also like to thank Linda L. Jensen of Brooks/Cole for her patient assistance.

Each of the programs has been carefully tested by the author. However, no express or implied warranty is made by the author or publisher in relation to the accuracy and functioning of the programs and related materials.

Allen H. Wolach

Contents

Chapter 1

Modifying Analysis of Variance Programs

The programs in this text have been tested on a variety of microcomputers. They can be run on most microcomputers without any modifications. A few simple changes in the programs can "tailor" them to a specific system.

Consider the Apple and Commodore (PET) microcomputer systems. The Apple uses a television set as a terminal monitor and the Commodore has an unusual built-in monitor. A standard terminal monitor can represent over 80 characters on one line of the monitor screen. The Apple or Commodore monitors can represent under 50 characters on a line of the monitor screen. Some of the statements in the present programs are over 50 characters in length. However, this is no problem for the Apple and Commodore microcomputers. The statements can be typed (entered) without modification. An Apple or Commodore microcomputer will automatically "reset" itself to the next line on the monitor screen when it has filled a line with characters. The listing displayed on the monitor screen will look like

```
10 PRINT "ENTER 1 TO CONTINUE ANALYSIS OF VARI
ANCE PROGRAM"
```

as opposed to

```
10 PRINT "ENTER 1 TO CONTINUE ANALYSIS OF VARIANCE PROGRAM"
```

A split line in the program listing will not alter the way the program runs or the output that the program generates on the monitor screen.

A more serious problem with the reduced number of characters per line for an Apple or Commodore microcomputer relates to program output. Consider Program 1 in Chapter 4. The output shown below is typical of the output for a microcomputer that displays 80 characters per line.

```
          SS             DF            MS            F
TOTAL     89.8169198     14
 BET.     42.2053986     2             21.1026993    5.31872087
 W.CEL    47.6115212     12            3.96762677
TOTAL     89.8169198     14
```

The Commodore microcomputer would produce the following output for the same program.

```
          SS             DF            MS
         F
TOTAL     89.8169198     14
 BET.     42.2053986     2             21.1
026993      5.31872087
 W.CEL    47.6115212     12            3.96
762677
TOTAL     89.8169198     14
```

Output from an Apple microcomputer would be similar to the output from the Commodore microcomputer. Note that the Apple and Commodore microcomputers do produce all of the program output. However, they destroy the tabled format of the output. The column headings SS, DF, MS, and F do not appear on one line of the monitor screen. The mean square of 21.1026993 is split into two parts, 21.1 and 026993.

Although the programs in Chapter 4 can be used in Apple or Commodore microcomputers without modification, the user may choose to modify the output format. First locate the offending output statements. They will all start with the word PRINT. For example, statement 770 in Program 1 is

```
770 PRINT "TOTAL";TAB(9);S1;TAB(22);N - 1
```

Any letters or numbers that occur inside of quotation marks are printed exactly as they are represented in the quotation marks.

Any variable such as S1 that is not inside of quotation marks will cause the current value of the variable to be printed out. Any expression such as N - 1 that is not inside of quotation marks will cause the value of the expression to be printed out. A TAB command such as TAB(38) will cause the microcomputer to start printing a given number of spaces from the left margin. TAB(38) starts printing 38 spaces from the left margin. Consider the statement

```
770 PRINT "TOTAL";TAB(9);S1;TAB(22);N - 1
```

The entire PRINT statement will be printed on one line. Statement 770 would produce the following kind of output

```
TOTAL    89.8169198    14
```

PRINT statement 770 causes TOTAL to be printed out because TOTAL is in quotation marks. The semicolon after "TOTAL" indicates that TAB(9) is distinct from "TOTAL." The TAB(9) command causes the microcomputer to move to the ninth space on the line. The semicolon after TAB(9) causes the microcomputer to print N - 1 starting at the ninth position on the line. Note that the value of N - 1, 14 in the present example, is printed because N - 1 is not inside of quotation marks. If N - 1 had been inside of quotation marks, the microcomputer would have literally printed N - 1.

Now we will examine the statements in Program 1 that print out the analysis of variance summary table. The relevant statements in the original program are

```
760 PRINT TAB(9);"SS";TAB(22);"DF";TAB(35);"MS";TAB(48);"F"
770 PRINT "TOTAL";TAB(9);S1;TAB(22);N - 1
780 PRINT " BET.";TAB(9);S2;TAB(22);T - 1;TAB(35);S2/(T - 1);
790 PRINT TAB(48);(S2/(T - 1))/(S3/(N - T))
800 PRINT " W.CEL";TAB(9);S3;TAB(22);N - T;TAB(35);S3/(N - T)
810 PRINT "TOTAL";TAB(9);S2 + S3;TAB(22);N - 1
```

We will examine statements 760 and 780 in detail. Assume that the statements are printed on a terminal monitor with at least 80 characters per line. The TAB(9) at the beginning of statement 760 starts printing at the ninth space on the line. Since SS is in quotation marks, SS is printed in the ninth and tenth positions on the line. Following the semicolon the TAB(22) command positions the internal microcomputer cursor to start printing at position 22 on the line. The "DF" that follows the semicolon is printed in positions 22 and 23 of the output line. TAB(35) positions the internal cursor to print MS in positions 35 and 36. Finally TAB(48) positions the cursor to print F in position 48. Note that no semicolon appears at the end of line 760. This insures that the next PRINT statement, statement 770, will start output on a new line of the monitor screen.

Now consider statement 780. This statement starts with " BET." The characters in the quotation marks are printed starting in the first space on the monitor line. Since the first character in the quotation marks is a space, the B in BET. starts at the second position on the line. The semicolon after " BET." causes the microcomputer to evaluate TAB(9) and move to the ninth position on the line. The value for S2 (42.2053986) is printed starting in the ninth position on the line. The number is printed because S2 is not in quotation marks. TAB(22) positions the cursor to print the value for T - 1 in position 22. TAB(35) moves the cursor to line position 35 so that the value for S2/(T - 1) can be printed. Note that S2/(T - 1) is followed by a semicolon. This semicolon insures that the next PRINT statement (statement 790) will continue to produce output on the same line of the monitor screen as was used by statement 780.

Now we shall rewrite the analysis of variance summary table PRINT statements for Program 1 so that an Apple or Commodore microcomputer can represent them with an acceptable format. The modified statements are:

```
760 PRINT TAB(9);"SS";TAB(22);"DF"
770 PRINT "TOTAL";TAB(9);S1;TAB(22);N - 1
780 PRINT " BET.";TAB(9);S2;TAB(22);T - 1
800 PRINT " W.CEL";TAB(9);S3;TAB(22);N - T
810 PRINT "TOTAL";TAB(9);S2 + S3
812 PRINT
```

```
814 PRINT TAB(9);"MS";TAB(22);"F"
816 PRINT " BET.";TAB(9);S2/(T - 1);
818 PRINT TAB(22);(S2/(T - 1))/(S3/(N - 1))
819 PRINT " W.CEL";TAB(9);S3/(N - T)
```

The modified program would produce the following output:

```
         SS            DF
TOTAL    89.8169198    14
 BET.    42.2053986    2
 W.CEL   47.6115212    12
TOTAL    89.8169198    14

         MS            F
 BET.    21.1026993    5.31872087
 W.CEL   3.96762677
```

The output shown above could easily be displayed on the monitor screen of an Apple or Commodore microcomputer. The sums of squares and degrees of freedom are printed first. Then the mean squares and F-ratios are printed. Row headings (BET., W.CEL) are repeated for the MS and F data to facilitate reading the output.

Let us consider the example of output for an Apple or Commodore computer in more detail. Statement 760 prints the column headings SS and DF. SS is started in column 9 because space must be reserved for the row headings in subsequent lines of the table. Statement 770 prints the row heading TOTAL and the values for the sum of squares total, S1, and the degrees of freedom total, N - 1. Statement 780 prints the row heading BET., the sum of squares between groups, and degrees of freedom between groups. The mean square and the F-ratio between groups are not printed out at this point. After all the sums of squares and degrees of freedom are printed out, statement 812 is used to place a space between the first part of the table and the beginning of the next part of the table. Then statement 814 prints out the column headings MS and F. Statements 816 and 818 print out the mean square and F ratio between groups. One could have entered statements 816 and 818 as the single statement

```
816 PRINT " BET.";TAB(9);S2/(T - 1);TAB(22);(S2/(T -1))/(S3/(N - T))
```

Finally, statement 819 is used to print out the within cell mean square.

If the user of this text does not think that he can modify the output of the programs, he can easily learn to read tables with improper format. Most microcmputer and minicomputer systems do not require modifying the format for analysis of variance summary tables.

Chapter 2

Alternate Names for Each Kind of Analysis of Variance

Sometimes it is difficult to recognize a kind of analysis of variance because one learns to identify the analysis with a different name. This chapter presents alternative names for analyses of variance. Pages in representative texts that describe each analysis of variance are also included. A complete reference for each text is in the reference list at the end of this book.

Programs 4, 5, and 6 in the present text are combined programs that can be used to perform more than one kind of analysis of variance. Programs 7 through 15 are used to perform only one kind of analysis of variance. Programs 1 through 3 can be used when one has unequal numbers of subjects for the conditions in an analysis of variance. Programs 4 through 17 require an equal number of subjects in each experimental condition.

Consider the following example. Programs 1, 4, and 7 are described as programs for an independent groups one-way analysis of variance. Program 1 is used when all of the groups for the analysis of variance do not have the same number of subjects. Of course, Program 1 can also be used if all groups have the same number of subjects. Program 4 is a combined program that can perform more than one kind of analysis of variance. An independent groups one-way analysis of variance is one of the analyses that can be performed with Program 4. Remember that the combined programs (Programs 4, 5, and 6) and Programs 7 through 17 require an equal number of observations in each cell.

The analysis of variance programs are presented in Chapter 4. Procedures for entering data are presented in Chapter 3.

Programs 1, 4, and 7

Independent groups one-way analysis of variance (completely randomized analysis of variance, CR-K)

Edwards (1972). Randomized group design; 113-119, 121.

Guilford (1978). Analysis in a one-way classification problem; 223-234, unequal n 234-235.

Hays (1981). A simple, one-way, analysis of variance; 333-350; 379-386.

Kirk (1982). Completely randomized; CR-P; 9-12, 139-142.

Loftus (1982). Analysis of variance; 311-329, 331-339, unequal n 329-330.

Winer (1971). Completely randomized design; 149-170.

Programs 4, and 8

Two-way analysis of variance without replications (randomized block, RB-K)

Edwards (1972). Randomized block design; 231-232.

Guilford (1978). A two-way classification analysis without replications; 268-270.

Hays (1981). Randomized blocks; 401-404.

Kirk (1982). Randomized block design, RB-P; 12-13, 243-245.

Loftus (1982). Within-subjects for repeated measures designs; 387-400.

Winer (1971). A two-way classification analysis without replications; 261-269.

Programs 2, 5, and 9

Independent groups two-way analysis of variance (completely randomized factorial analysis of variance, CRF-JK)

Edwards (1972). Randomized groups analysis of variance with each cell as a single observation (the procedure can provide an unweighted means solution when the number of observations in each cell are not equal); 217-220: factorial experiments with unequal n´s; 217-220: using treatment means as single observations; 217-220.

Guilford (1978). Analysis in a two-way classfication problem; 242-257.

Hays (1981). Two-way analysis of variance with replications; 353-365; 387-394; 394-399.

Kirk (1982). Completely randomized factorial design, CRF-pq; 16-18; 353-357.

Loftus (1982). Two-way analysis of variance; 344-375.

Winer (1971). Factorial experiment having n observations per cell--pxq; 431-440: factorial experiment pxq, unequal n; 445-452.

Programs 5, and 10

Split-plot analysis of variance, SPF-J.K

Edwards (1972). Split-plot design; 276-280.

Kirk (1982). Split-plot design, factorial design with block-treatment confounding, SPF-p.q; 492-496.

Winer (1971). Two factor experiment with repeated measures on one factor; 518-526.

Programs 5, and 11

Randomized block factorial analysis of variance, RBF-.JK

Edwards (1972). Factorial experiment with a randomized block design; 240-244*.

Kirk (1982). Randomized block factorial design, RBF-pq; 443-445*.

Loftus (1982). Multiple observations per subject per condition; 400-410**

Winer (1971). Other multifactor plans; 575-576

*The following table shows what the present programs print out, Edward´s notation, and Kirk´s notation.

Present programs	Edwards	Kirk
BLOCKS	B:BLOCKS	BLOCKS
A	A	A
AXBLOCKS	BLOCKXA	AXBLOCKS
B	C	B
BXBLOCKS	BLOCKXC	BXBLOCKS
AB	AC	AB
ABXBLOCKS	BLOCKXAXC	ABXBLOCKS

Note that the sums of squares for AXBLOCKS, BXBLOCKS, and ABXBLOCKS can be summed to form a composite residual error term. Similarly the degrees of freedom for AXBLOCKS, BXBLOCKS, and ABXBLOCKS, can be summed to form degrees of freedom for the residual term. The following table shows how the present programs can have error terms combined. The table also shows how Edwards and Kirk would represent an analysis of variance with a combined residual term.

Present programs	Edwards	Kirk
BLOCKS	B:BLOCKS	BLOCKS
A	A	A
B	C	B
AB	AC	AB
RESIDUAL (AXBLOCKS + BXBLOCKS + ABXBLOCKS)	BXT (BLOCKXA + BLOCKXC + BLOCKXAXC	RESIDUAL (AXBLOCKS + BXBLOCKS + ABXBLOCKS)

**The randomized block factorial design program can be used to calculate terms for what Loftus and Loftus refer to as a multiple observations per subject per condition design. The row sum of squares in their text corresponds to the blocks sum of squares in the present programs. The column sum of squares corresponds to the Factor B sum of squares in the present programs. The interaction sum of squares that Loftus and Loftus refer to is the BXBLOCKS sum of squares in the present programs. Their within sum of squares is the sum of the sums of squares A, AXBLOCKS, and ABXBLOCKS in the present programs.

Programs 3, 6, and 12

Independent groups three-way analysis of variance (completely randomized factorial analysis of variance, CRF-JKL)

Guilford (1978). A three-way classification factorial design; 258-267.

Kirk (1982). Completely randomized factorial design, CRF-pqr; 431-435.

Loftus (1982). Higher-order ANOVA; A three-way ANOVA; 375-378.

Winer (1971). pxqxr factorial experiment having n observations per cell; 452-464.

Programs 6, and 13

Split-plot analysis of variance (one repeated measure, SPF-JK.L)

Kirk (1982). Split plot design, SPF-pr.q; 524-527*.

Winer (1971). Three factor experiment with repeated measures (case 2); 559-567*.

*Kirk makes Factor B (the second factor) the repeated measures factor. The programs in the present text and Winer's text make Factor C (the third factor) the repeated measures factor. The table shows the source table for the split-plot analysis of variance, SPF-JK.L, in the present programs, Kirk's text, and Winer's text.

Present programs	Kirk's text	Winer's text
BETWEEN SUBJECTS	BETWEEN BLOCKS	BETWEEN SUBJECTS
A	A	A
B	C	B
AB	AC	AB
SUBJ. W. GROUPS	BLOCKS. W. AC	SUBJ. W. GROUPS
WITHIN SUBJECTS	WITHIN BLOCKS	WITHIN SUBJECTS
C	B	C
AC	AB	AC
BC	BC	BC
ABC	ABC	ABC
CXSUBJ. W. GROUPS	BLOCKS W. AC	CXSUBJ. W. GROUPS

Programs 6, and 14

Split-plot analysis of variance (two repeated measures, SPF-J.KL)

Kirk (1982). Split-plot design, SPF-p.qr; 535-540.

Winer (1971). Three factor experiment with repeated measures (case 1); 539-550.

Programs 6, and 15

Randomized block factorial analysis of variance (RBF-.JKL)

None of the texts include an explicit explanation of this design.

Program 16

Latin square design, LS-K

Edwards (1972). Latin square design; 285-300.

Kirk (1982). Latin square, LS-P; 13-16, 312-317.

Winer (1971). Latin squares--no repeated measures; 685-701.

Program 17

Analysis of covariance for a randomized groups design

Edwards (1972). Analysis of covariance, analysis of covariance for a randomized group design; 369-388.

Kirk (1982). Analysis of covariance, CRAC-P; 728-730.

Winer (1971). Analysis of covariance, a single factor experiment; 775-780.

Chapter 3

How to Enter Data in Analysis of Variance Programs

Make sure that the microcomputer is set for programming in BASIC. Select the appropriate program from Chapter 2 and enter it statement by statement. Once the program is entered it can be recorded on tape (or disk) so that it does not have to be hand entered every time it is used.

Data are entered in DATA statements starting with DATA statement 5000. The number of numbers (parameters or observations) per statement is not critical. However, data must be entered in the appropriate order in DATA statements that have ascending statement numbers. The statement numbers do not have to be consecutive. Consider the following examples for entering the numbers 1, 3, 5, 25, 2, 200, 3, 4, 4000, 57, 62, 54.

Example 1

```
5000 DATA 1, 3, 5, 25
5001 DATA 2, 200, 3, 4
5002 DATA 4000, 57, 62, 54
```

Example 2

```
5000 DATA 1, 3, 5, 25
5010 DATA 2, 200, 3, 4
5020 DATA 4000, 57, 62, 54
```

Note that ascending DATA statements in both examples contain the same numbers. The statement numbers are consecutive in Example 1, but not in Example 2. Both methods for entering data successfully enter the data for the analysis of variance program.

Most versions of BASIC require the highest numbered statement to be no greater than 9999. The following information should be useful for creating DATA statements:

1. The word DATA must be typed (entered) after each statement number.
2. Each number in a DATA statement is separated from the next number by a comma.
3. A comma is not used before the first number or after the last number in a DATA statement.
4. Spaces before or after the key word DATA or before or after a comma are optional.
5. Spaces cannot occur within a number; e.g., 25 is not the same as 2 5.
6. Numbers in a DATA statement are assumed to be positive when they are not preceded by a sign.
7. A negative number is preceded by a negative (-) sign.
8. Decimals can be inserted in numbers.
9. Numbers can be expressed in exponential notation. The observations for the present analysis of variance programs should not be large enough to require scientific (exponential) notation. Exponential calculations may produce inaccurate results that are beyond the nine digit capacity of the BASIC interpreter for a microcomputer. The difference formulas used in the present programs minimize the possibility that the calculations will require scientific notation.

Consider the following DATA statement:

```
5000 DATA 5, 5.23, -2.1, -2
```

This data statement contains the numbers +5, + 5.23, -2.1, and -2.

Programs that include a sum of squares total (Programs 4 through 15) have the sum of squares calculated twice, once from the original observations, and once by summing the sums of squares that are supposed to add up to the sum of squares total. If the two sums of squares for the sum of squares total are similar numbers (only differ in the last few decimal places), one knows that the computer system accurately performed the analysis of variance.

Most microcomputer BASIC interpreters have some editing capabilities. One can delete a DATA statement by typing (entering) the statement number followed by a carriage return. One can change a DATA statement by depressing the carriage return, entering the statement number, and then entering the new statement. The order in which DATA statements are entered is not critical. DATA statement 5020 can be entered before DATA statement 5010. However, if a number occurs after another number in the data list, it must occur in a higher numbered DATA statement, or later in the same DATA statement.

Once the data for a program are entered, the program is started by typing RUN followed by a carriage return. The program will stop before the monitor overflows (has more data than it can display on the screen). The monitor will cue the user about continuing the program after the user transcribes the data. For example, if the monitor indicates ENTER 1 FOLLOWED BY A CARRIAGE RETURN TO CONTINUE, the user enters the number 1 and then depresses the carriage return key.

The program can be rerun by entering RUN followed by a carriage return. If the user wants to use the program with a new set of observations, she must remove all data from the previous program. This can be accomplished by entering the statement number followed by a carriage return. This process must be repeated for each DATA statement that was in the previous analysis. Note that the second to last statement in most of the programs is a DATA statement. This statement must not be removed from the program. Also note that this DATA statement has a higher statement number than all of the DATA statements generated by the user.

Saving Programs

Most microcomputer systems with BASIC allow the user to save a program on tape with a command such as SAVE followed by a carriage return. The user can save the program before the observations are entered. If the user wants to retain the data with the program, he can SAVE the program after the observations are entered. Microcomputers have commands for listing program statements on the monitor. Entering LIST 5000-5030, followed by a carriage return will produce a listing of statements 5000 through 5030.

Data Entry

The following pages contain an example of data entry for each program in this text. Statement 5000 will always be used to specify the parameters for the analysis of variance (e.g, number of levels for each factor).

Program 1. Completely randomized analysis of variance (does not require equal n in all cells), CR-K

DATA statement 5000 should have one number, the number of treatment groups. For example, if there are 7 treatment groups one would enter:

5000 DATA 7

The data are then entered in ascending DATA statements, one treatment group at a time; i.e., all of the observations for Treatment Group 1 before all observations for Treatment Group 2, etc. After observations for a given treatment group are entered the number 3E23 is entered before the first observation is entered for the next group. The program uses the number 3E23 to find the end of the data for a given group. Do not forget to enter 3E23 after the last observation for the last group.

Data for an independent groups one-way analysis of variance, CR-K are given below. An actual analysis of variance would usually have more observations per treatment group.

Treatment Group 1	Treatment Group 2	Treatment Group 3	Treatment Group 4
5.2	7.3	2.1	15.1
6.1	9.1	4.0	16.2
	9.1	3.0	15.1
	10.4		14.3
			15.1

The data for the analysis of variance could be entered in DATA statements as follows:

```
5000 DATA 4
5010 DATA 5.2, 6.1, 3E23, 7.3, 9.1, 9.1, 10.4, 3E23
5020 DATA 2.1, 4.0, 3.0, 3E23, 15.1, 16.2, 15.1
5030 DATA 14.3, 15.1, 3E23
```

When the program is run with the observations shown above, it should supply the following information:

Number of subjects per treatment group

N1 = 2

N2 = 4

N3 = 3

N4 = 5

	Treatment Group 1	Treatment Group 2	Treatment Group 3	Treatment Group 4
Mean	5.65	8.975	3.03333333	15.16
Variance	.405	1.62249999	.903333332	.458

n - 1 is used in the denominator of the variance formula.

	SS	DF	MS	F
TOTAL	325.232139	13		
BET.	316.320974	3	105.440324	118.323815
W.CEL	8.91116666	10	0.891116666	
TOTAL	325.23214	13		

Program 2. Completely randomized factorial analysis of variance (does not require equal n in all cells), CRF-JK

Data statement 5000 should have two numbers, the number of levels of Factor A (the first factor), and the number of levels of Factor B (the second factor). Consider the following example:

5000 DATA 3, 5

The example indicates that Factor A has 3 levels, and Factor B has 5 levels.

All the observations for one subgroup must be entered before the observations for the next subgroup. After the last observation for a subgroup is entered the number 3E23 is entered to indicate the end of the data for the subgroup.

The subgroups are entered with data from all subgroups at the first level of Factor A entered before all subgroups at the second level of Factor A, etc. The subgroups at a given level of Factor A are always entered with the levels of Factor B entered in the same order. Suppose that one has an analysis of variance with 3 levels of Factor A and 5 levels of Factor B. The subgroups would be entered in the order A1B1, A1B2, A1B3, A1B4, A1B5, A2B1, A2B2, A2B3, A2B4, A2B5, A3B1, A3B2, A3B3, A3B4, A3B5. Observation summary tables for two independent groups two-way analyses of variance are shown below:

Factor B

	B1	B2	B3	B4	B5
A1	A1B1 *1*	A1B2 *2*	A1B3 *3*	A1B4 *4*	A1B5 *5*
A2	A2B1 *6*	A2B2 *7*	A2B3 *8*	A2B4 *9*	A2B5 *10*
A3	A3B1 *11*	A3B2 *12*	A3B3 *13*	A3B4 *14*	A3B5 *15*

	Factor B		
	B1	B2	B3
A1	A1B1 *1*	A1B2 *2*	A1B3 *3*
A2	A2B1 *4*	A2B2 *5*	A2B3 *6*
A3	A3B1 *7*	A3B2 *8*	A3B3 *9*
A4	A4B1 *10*	A4B2 *11*	A4B3 *12*

The italicized numbers in the examples show the order in which observations from subgroups are entered in the analysis of variance program. All the observations in cell 1 (A1B1) for a given analysis would be entered before the observations in cell 2, etc.

Data for an independent groups two-way analysis of variance are given below. An actual analysis of variance would usually have more observations per subgroup.

	Factor B		
	B1	B2	B3
A1	7.93	7.39	3.01
	7.22	6.05	3.36
		6.57	5.83
		5.01	7.04
		4.26	7.12
			3.07
A2	7.03	6.82	4.85
	8.48	7.10	4.09
		4.42	6.11
			7.97

The data for the analysis of variance could be entered in DATA statements as follows:

```
5000 DATA 2, 3
5010 DATA 7.93, 7.22, 3E23
5020 DATA 7.39, 6.05, 6.57, 5.01, 4.26, 3E23
5030 DATA 3.01, 3.36, 5.83, 7.04, 7.12, 3.07, 3E23
5040 DATA 7.03, 8.48, 3E23, 6.82, 7.10, 4.42, 3E23
5050 DATA 4.85, 4.09, 6.11, 7.97, 3E23
```

When the program is run with the observations shown above, it should supply the following information:

```
Number of subjects per subgroup
        N for A1B1 = 2
        N for A1B2 = 5
        N for A1B3 = 6
        N for A2B1 = 2
        N for A2B2 = 3
        N for A2B3 = 4
```

Factor B

	B1	B2	B3
A1	M11 = 7.575 V11 = .25205	M12 = 5.856 V12 = 1.54088	M13 = 4.90499997 V13 = 3.933069995
A2	M21 = 7.755 V21 = 1.05125	M22 = 6.11333332 V22 = 2.17013332	M23 = 5.755 V23 = 2.87449999

Note that Mjk is the mean for a cell, and Vjk is the variance for a cell with n - 1 degrees of freedom in the denominator of the variance formula.

Means for levels of Factor A

A1 = 6.11199999

A2 = 6.5411111

Means for levels of Factor B

B1 = 7.665

B2 = 5.98466666

B3 = 5.32999998

	SS	DF	MS	F
A	0.276204515	1	0.276204515	0.33913065
B	5.80288908	2	2.90144454	3.56246448
AB	0.13435568	2	0.06717784	0.0824825929
WITH		16	0.814448693	

DATA statement 5000 should have three numbers, the number of levels of Factor A (the first factor), the number of levels of Factor B (the second factor), and the number of levels of Factor C (the third factor). Consider the following example:

```
5000 DATA 4, 5, 7
```

The example indicates that Factor A has 4 levels, Factor B has 5 levels, and Factor C has 7 levels.

All the observations for one subgroup must be entered before the observations for the next subgroup. After the last observation for a subgroup is entered the number 3E23 is entered to indicate the end of the data for the subgroup.

Observations are entered with all the subgroups at the first level of Factor A and the first level of Factor B entered first; i.e., A1B1C1, A1B1C2, A1B1C3 ... A1B1Cn. Then observations are entered at the first level of Factor A and the second level of Factor B; i.e., A1B2C1, A1B2C2, A1B2C3 ... A1B2Cn. This procedure is continued until all the subgroups at level A1 are entered. Then the subgroups at level A2 are entered. First subgroups A2B1C1, A2B1C2, A2B1C3 ... A2B1Cn are entered. Then subgroups A2B2C1, A2B2C2, A2B2C3 ... A2B2Cn are entered.

The italicized numbers in the examples show the order in which observations from subgroups are entered in the analysis of variance program. All of the observations for cell 1 in a given analysis would be entered before the observations in cell 2, etc. Observation summary tables for two independent groups three-way analyses of variance are shown below:

		C1	C2	C3	C4	C5
A1	B1	1	2	3	4	5
	B2	6	7	8	9	10
	B3	11	12	13	14	15
	B4	16	17	18	19	20
A2	B1	21	22	23	24	25
	B2	26	27	28	29	30
	B3	31	32	33	34	35
	B4	36	37	38	39	40
A3	B1	41	42	43	44	45
	B2	46	47	48	49	50
	B3	51	52	53	54	55
	B4	56	57	58	59	60

		C1	C2
A1	B1	1	2
	B2	3	4
	B3	5	6
	B4	7	8
	B5	9	10
A2	B1	11	12
	B2	13	14
	B3	15	16
	B4	17	18
	B5	19	20
A3	B1	21	22
	B2	23	24
	B3	25	26
	B4	27	28
	B5	29	30

Data for an independent groups three-way analysis of variance are given below. An actual analysis of variance would usually have more observations per subgroup.

		C1	C2	C3	C4
A1	B1	4.56 6.42 3.25	9.79 11.49	9.23 11.56 8.54 9.16	9.01 9.36 11.83
	B2	6.15 1.94	6.48 5.94 9.22 6.78	11.52 8.78 6.48	13.04 13.13
	B3	4.44 5.67 5.25 5.83	7.89 4.01	8.61 9.53	9.07 13.03 14.49 12.66
A2	B1	8.47 5.25 4.90	9.27 11.82 9.15	12.27 14.82	15.48 16.27 14.52
	B2	8.80 4.79 6.43	8.98 10.59	12.15 11.98 13.59	13.46 17.70 12.64 14.18
	B3	7.73 7.73 4.04 7.18	8.80 9.87 8.41	11.80 12.87	12.32 14.30

The data for the analysis of variance could be entered in DATA statements as follows:

```
5000 DATA 2,3,4
5010 DATA 4.56,6.42,3.25,3E23,9.79,11.49,3E23
5020 DATA 9.23,11.56,8.54,9.16,3E23,9.01,9.36,11.83,3E23
5030 DATA 6.15,1.94,3E23,6.48,5.94,9.22,6.78,3E23
5040 DATA 11.52,8.78,6.48,3E23,13.04,13.13,3E23
5050 DATA 4.44,5.67,5.25,5.83,3E23,7.89,4.01,3E23
5060 DATA 8.61,9.53,3E23,9.07,13.03,14.49,12.66,3E23
5070 DATA 8.47,5.25,4.90,3E23,9.27,11.82,9.15,3E23
5080 DATA 12.27,14.82,3E23,15.48,16.27,14.52,3E23
5090 DATA 8.80,4.79,6.43,3E23,8.98,10.59,3E23
5100 DATA 12.15,11.98,13.59,3E23,13.46,17.70,12.64,14.18,3E23
5110 DATA 7.73,7.73,4.04,7.18,3E23,8.80,9.87,8.41,3E23
5120 DATA 11.80,12.87,3E23,12.32,14.30,3E23
```

When the program is run with the above data it should supply the following information:

```
Number of subjects per subgroup
        N for A1B1C1 = 3
        N for A1B1C2 = 2
        N for A1B1C3 = 4
        N for A1B1C4 = 3
        N for A1B2C1 = 2
        N for A1B2C2 = 4
        N for A1B2C3 = 3
        N for A1B2C4 = 2
        N for A1B3C1 = 4
        N for A1B3C2 = 2
        N for A1B3C3 = 2
        N for A1B3C4 = 4
        N for A2B1C1 = 3
        N for A2B1C2 = 3
        N for A2B1C3 = 2
        N for A2B1C4 = 3
        N for A2B2C1 = 3
```

N for A2B2C2 = 2

N for A2B2C3 = 3

N for A2B2C4 = 4

N for A2B3C1 = 4

N for A2B3C2 = 3

N for A2B3C3 = 2

N for A2B3C4 = 2

		C1	C2	C3	C4
A1	B1	M111 = 4.74333333 V111 = 2.5374333	M112 = 10.64 V112 = 1.445	M113 = 9.6225 V113 = 1.76455832	M114 = 10.0666666 V114 = 2.36263331
	B2	M121 = 4.045 V121 = 8.86205	M122 = 7.105 V122 = 2.10889999	M123 = 8.92666666 V123 = 6.36653331	M124 = 13.085 V124 = .00405
	B3	M131 = 5.2975 V131 = .386624999	M132 = 5.95 V132 = 7.5272	M133 = 9.07 V133 = .4232	M134 = 12.3125 V134 = 5.2969583
A2	B1	M211 = 6.20666666 V211 = 3.872633	M212 = 10.08 V212 = 2.2743	M213 = 13.545 V213 = 3.25125	M214 = 15.4233333 V214 = .768033332
	B2	M221 = 6.67333332 V221 = 4.06443332	M222 = 9.785 V222 = 1.29605	M223 = 12.5733333 V223 = .782433318	M224 = 14.495 V224 = 4.96116666
	B3	M231 = 6.67 V231 = 3.14139999	M232 = 9.02666666 V232 = .571433332	M233 = 12.335 V233 = .57245	M234 = 13.31 V234 = 1.9602

Note that Mjkl is the mean for a cell and Vjkl is the variance for a cell with n - 1 degrees of freedom in the denominator of the variance formula.

Means for AB interaction

	B1	B2	B3
A1	8.76812498	8.29041666	8.1575
A2	11.3137499	10.8816666	10.3354166

Means for AC interaction

	C1	C2	C3	C4
A1	4.69527777	7.89833332	9.20638888	11.8213888
A2	6.51666665	9.63055554	12.8177777	14.4094444

Means for BC interaction

	C1	C2	C3	C4
B1	5.47499999	10.36	11.58375	12.7449999
B2	5.35916666	8.445	10.7499999	13.79
B3	5.98375	7.48833333	10.7025	12.81125

Means for levels of Factor A

A1 = 8.4053472

A2 = 10.843611

Means for levels of Factor B

B1 = 10.0409374

B2 = 9.58604163

B3 = 9.24645833

Means for levels of Factor C

C1 = 5.6059722

C2 = 8.76444443

C3 = 11.0120832

C4 = 13.1154166

	SS	DF	MS	F
A	35.6707818	1	35.6707818	34.102085
B	2.54251725	2	1.27125862	1.21535238
C	186.00088	3	62.0002933	59.2737018
AB	0.20542445	2	0.102712225	0.0981952419
AC	3.416535	3	1.138845	1.08876192
BC	8.803894	6	1.46731566	1.40278741
ABC	10.1655085	6	1.69425141	1.61974319
W.CEL		45	1.04600002	

Programs 4 and 7. Completely randomized analysis of variance (requires equal n in all cells), CR-K

When these programs are used for an independent groups one-way analysis of variance (CR – K), DATA statement 5000 should have three numbers. The first number is a 0 which indicates there are no repeated measures in this analysis of variance. The second number indicates the number of subjects per treatment group, and the third number indicates the number of treatment groups. These programs require the same number of subjects in each treatment group. For example, if one had 4 treatment groups with 9 subjects per group, the DATA statement would be as follows:

```
5000 Data 0, 9, 4
```

The observations are then entered in ascending DATA statements, one treatment group at a time; i.e., all the observations for Treatment Group 1 before all observations for Treatment Group 2, etc.

Data for an independent groups one-way analysis of variance are given below. An actual analysis of variance would usually have more observations.

Treatment Group 1	Treatment Group 2	Treatment Group 3
3.012	3.071	7.012
3.363	7.032	7.363
5.832	8.485	9.832
7.043	6.658	11.043
7.128	7.065	11.128

The data for the analysis of variance could be entered in DATA statements as follows:

```
5000 DATA 0, 5, 3
5010 DATA 3.012, 3.363, 5.832, 7.043, 7.128
```

```
5020 DATA 3.071, 7.032, 8.485, 6.658, 7.065
5030 DATA 7.012, 7.363, 9.832, 11.043, 11.128
```

When the program is run with the observations shown above, it should supply the following information:

	Treatment Group 1	Treatment Group 2	Treatment Group 3
Mean	5.2756	6.4622	9.2756
Variance	3.9116483	4.07958369	3.9116483

n - 1 is used in the denominator of the variance formula.

	SS	DF	MS	F
TOTAL	89.8169188	14		
TREAT.	42.2053982	2	21.1026991	5.31872089
W. CEL	47.6115206	12	3.96762671	
TOTAL	89.8169188	14		

Programs 4 and 8. Randomized block analysis of variance (requires equal n in all cells), RB-K

When these programs are used for a two-way analysis of variance without replications (RB-K), DATA statement 5000 should have three numbers. The first number is a 1 which indicates there is one repeated measure (block) factor in this analysis of variance and the second number indicates the number of blocks. The third number indicates the number of treatments. Suppose one had 5 treatments with 7 blocks; i.e., one observation from each block at each treatment level. The appropriate DATA statement for an experiment with 5 treatment levels and 7 blocks is:

```
5000 DATA 1, 7, 5
```

The observations are then entered in ascending DATA statements, one treatment at a time; i.e., all the observations for Treatment 1 are entered before all observations for Treatment 2, etc. The blocks must be entered in the same order for each treatment.

Data for a two-way analysis of variance without replications are given below. An actual analysis of variance would usually have more blocks.

Blocks	Treatment 1	Treatment 2	Treatment 3	Treatment 4	Treatment 5
1	8.01	7.01	7.28	5.01	4.01
2	8.36	7.36	6.23	5.36	6.83
3	10.83	9.83	9.04	7.83	4.36
4	12.04	11.04	9.13	9.04	8.14

The data for the analysis of variance could be entered in DATA statements as follows:

```
5000 DATA 1, 4, 5
```

```
5010 DATA 8.01, 8.36, 10.83, 12.04
5020 DATA 7.01, 7.36, 9.83, 11.04
5030 DATA 7.28, 6.23, 9.04, 9.13
5040 DATA 5.01, 5.36, 7.83, 9.04
5050 DATA 4.01, 6.83, 4.36, 8.14
```

When the program is run with the observations shown above, it should supply the following information:

	Treatment 1	Treatment 2	Treatment 3	Treatment 4	Treatment 5
Mean	9.81	8.81	7.92	6.81	5.835
Var.	3.78526666	3.78526666	1.99473332	3.78526666	3.9364332

The denominator of each variance calculation is the number of blocks minus 1 (n - 1).

	Block 1	Block 2	Block 3	Block 4
Mean	6.264	6.828	8.378	9.878
Var.	2.82258	1.28567	6.25087	2.57672

The denominator of each variance calculation is the number of treatments minus 1 (J - 1).

	SS	DF	MS	F
TOTAL	91.49722	19		
BLOCKS	39.75386	3	13.2512866	13.1341301
TREAT.	39.63632	4	9.90908	9.82147246
REMAIN	12.10704	12	1.00892	
TOTAL	91.49722	19		

When these programs are used for an independent groups two-way analysis of variance (CRF-JK), DATA statement 5000 should have four numbers. The first number is a 0 which indicates there are no repeated measures (blocks) factors in the analysis of variance. The second number indicates the number of observations in a subgroup. The third number indicates the number of levels of Factor A (first factor), and the fourth number indicates the number of levels of Factor B (second factor). Consider the following example:

```
5000 DATA 0, 25, 9, 12
```

This example indicates that Factor A has 9 levels and Factor B has 12 levels. All of the subgroups have 25 observations. Remember that the number of observations for each subgroup must be the same for these programs.

All of the observations for one subgroup must be entered before the observations for the next subgroup. The subgroups are entered with data from all subgroups at the first level of Factor A entered before all subgroups at the second level of Factor A, etc. The subgroups at a given level of Factor A are always entered with the levels of Factor B entered in the same order. Suppose that one has an analysis of variance with 3 levels of Factor A and 4 levels of Factor B. The subgroups would be entered in the order A1B1, A1B2, A1B3, A1B4, A2B1, A2B2, A2B3, A2B4, A3B1, A3B2, A3B3, A3B4.

Observation summary tables for two independent groups two-way analyses of variance are shown below:

Factor B

	B1	B2	B3	B4
A1	A1B1 *1*	A1B2 *2*	A1B3 *3*	A1B4 *4*
A2	A2B1 *5*	A2B2 *6*	A2B3 *7*	A2B4 *8*

Factor B

	B1	B2
A1	A1B1 *1*	A1B2 *2*
A2	A2B1 *3*	A2B2 *4*
A3	A3B1 *5*	A3B2 *6*
A4	A4B1 *7*	A4B2 *8*

The italicized numbers in the examples show the order in which observations from cells are entered in the analysis of variance programs. All of the observations in cell 1 for a given analysis would be entered before the observations in cell 2, etc.

Data for an independent groups two-way analysis of variance are given below. An actual analysis of variance would usually have more observations per subgroup.

Factor B

	B1	B2	B3
A1	7.3 9.8	7.0 8.6	7.9 6.4
A2	5.8 6.1	8.0 5.7	5.3 4.8
A3	7.5 8.7	9.9 9.4	10.3 5.5
A4	2.9 2.0	2.4 5.0	0.7 1.7

When the program is run with the observations shown above, it should supply the following information:

Factor B

	B1	B2	B3
A1	M11 = 8.55 V11 = 3.125	M12 = 7.8 V12 = 1.28	M13 = 7.15 V13 = 1.25
A2	M21 = 5.95 V21 = .045	M22 = 6.85 V22 = 2.645	M23 = 5.05 V23 = .125
A3	M31 = 8.1 V31 = .72	M32 = 9.65 V32 = .125	M33 = 7.9 V33 = 11.52
A4	M41 = 2.45 V41 = .405	M42 = 3.7 V42 = 3.38	M43 = 1.2 V43 = .5

Note that Mjk is the mean for a cell, and Vjk is the variance for a cell with n - 1 degrees of freedom in the denominator of the variance formula.

The data for the analysis of variance could be entered in DATA statements as follows:

```
5000 DATA 0, 2, 4, 3
5010 DATA 7.3, 9.8, 7.0, 8.6, 7.9, 6.4
5020 DATA 5.8, 6.1, 8.0, 5.7, 5.3, 4.8
5030 DATA 7.5, 8.7, 9.9, 9.4, 10.3, 5.5
5040 DATA 2.9, 2.0, 2.4, 5.0, .7, 1.7
```

```
Means for levels of Factor A
     A1 = 7.8333333
     A2 = 5.94999998
     A3 = 8.54999998
     A4 = 2.44999998

Means for levels of Factor B
     B1 = 6.2625
     B2 = 7
     B3 = 5.325
```

	SS	DF	MS	F
TOTAL	174.009581	23		
A	133.891248	3	44.630416	21.4268851
B	11.2758332	2	5.6379166	2.70674132
AB	3.8475008	6	0.641250133	0.307861637
W.CEL	24.995	12	2.08291666	
TOTAL	174.009581	23		

Programs 5 and 10. Split-plot analysis of variance (requires equal n in all cells), SPF-J.K

When these programs are used for a split-plot analysis of variance (SPF-J.K), DATA statement 5000 should have four numbers. The first number is a 1 which indicates there is one repeated measures factor in the analysis of variance. The second number is the number of observations in each cell. The third number is the number of levels of Factor A, and the fourth number is the number of levels of Factor B. Factor B is the repeated measures factor. Remember that these programs require an equal number of observations in each cell. Consider the following example:

```
5000 DATA 1, 16, 8, 12
```

This example indicates that Factor A has 8 levels and Factor B, the repeated measures factor, has 12 levels. The analysis has 16 observations in each cell.

All the observations for one cell must be entered before the observations for the next cell. If a cell contains observations that are related to the observations in the preceding cells (repeated measures), related observations must be entered in the same order as in the preceding cell(s). The cells are entered with data from all cells at the first level of Factor A entered before all cells at the second level of Factor A, etc. The cells at a given level of Factor A are always entered with the levels of Factor B entered in the same order. Suppose that one has a split-plot analysis of variance with two levels of Factor A and three levels of Factor B. The cells would be entered in the order A1B1, A1B2, A1B3, A2B1, A2B2, A2B3. All the cells at a given level of Factor A have the related observations entered in the same order.

Observation summary tables for two split-plot (SPF-J.K) analyses of variance are shown below:

Factor B

		B1	B2	B3
		A1B1	A1B2	A1B3
	Block 1	*1*	*4*	*7*
A1	Block 2	*2*	*5*	*8*
	Block 3	*3*	*6*	*9*
		A2B1	A2B2	A2B3
	Block 4	*10*	*13*	*16*
A2	Block 5	*11*	*14*	*17*
	Block 6	*12*	*15*	*18*

Factor B

		B1	B2
		A1B1	A1B2
	Block 1	*1*	*5*
	Block 2	*2*	*6*
A1	Block 3	*3*	*7*
	Block 4	*4*	*8*
		A2B1	A2B2
	Block 5	*9*	*13*
	Block 6	*10*	*14*
A2	Block 7	*11*	*15*
	Block 8	*12*	*16*
		A3B1	A3B2
	Block 9	*17*	*21*
	Block 10	*18*	*22*
A3	Block 11	*19*	*23*
	Block 12	*20*	*24*

The italicized numbers in the examples show the order in

which observations are entered in the split-plot analysis of variance (SPF-J.K) programs.

Data for a split-plot analysis of variance (SPF-J.K) are given below. An actual analysis of variance would usually have more blocks.

Factor B

		B1	B2	B3	B4
A1	Block 1	938	1006	852	979
	Block 2	712	818	721	1240
	Block 3	592	838	831	1419
A2	Block 4	714	398	106	35
	Block 5	1104	340	377	29
	Block 6	854	351	437	484

The data for the analysis of variance could be entered in DATA statements as follows:

```
5000 DATA 1, 3, 2, 4
5010 DATA 938, 712, 592, 1006, 818, 838
5020 DATA 852, 721, 831, 979, 1240, 1419
5030 DATA 714, 1104, 854, 398, 340, 351
5040 DATA 106, 377, 437, 35, 29, 484
```

When the program is run with the observations shown above, it should supply the following information:

Factor B

	B1	B2	B3	B4
A1	M11 = 747.333332 V11 = 30865.333	M12 = 887.333332 V12 = 10661.3331	M13 = 801.333333 V13 = 4950.33332	M14 = 1212.6666 V14 = 48960.3331
A2	M21 = 890.666666 V21 = 39033.33	M22 = 362.999999 V22 = 948.999994	M23 = 306.666665 V23 = 31100.3331	M24 = 182.666666 V24 = 68110.3329

Note that Mjk is the mean for a cell, and Vjk is the variance for a cell with n - 1 degrees of freedom in the denominator of the variance formula.

Means for levels of Factor A

A1 = 912.166663
A2 = 435.749995

Means for levels of Factor B

B1 = 818.999998
B2 = 625.166664
B3 = 553.999998
B4 = 697.666664

Means for Blocks

Block 1 = 943.75
Block 2 = 872.75
Block 3 = 920
Block 4 = 313.25
Block 5 = 462.5
Block 6 = 531.5

	SS	DF	MS	F
TOTAL	3101076.89	23		
BET	1471846.7	5		
A	1361837.04	1	1361837.04	49.516998
SWG	110009.661	4	27502.4152	
WITH	1629230.19	18		
B	230218.737	3	76739.5956	2.56331967
AB	1039760.42	3	346586.806	11.5769802
BXSWG	359250.997	12	29937.583	
TOTAL	3101076.89	23		

Programs 5 and 11. Randomized block factorial analysis of variance (requires equal n in all cells), RBF-.JK

When these programs are used for a randomized block factorial analysis of variance (RBF-.JK), DATA statement 5000 should have four numbers. The first number is a 2 which indicates Factors A and B are repeated measures factors. The second number is the number of observations in each cell. The third number is the number of levels of Factor A, and the fourth number is the number of levels of Factor B. Remember that these programs require an equal number of observations in each cell. Consider the following example:

```
5000 DATA 2, 9, 6, 7
```

The example indicates that Factor A has 6 levels and Factor B has 7 levels. The analysis has 9 observations in each cell.

All the observations for one cell must be entered before the observations for the next cell. The blocks in a given cell must be entered in the same order that they are entered in all the other cells. The cells are entered with data from all cells at the first level of Factor A entered before all cells at the second level of Factor A, etc. The cells at a given level of Factor A are always entered with the levels of Factor B entered in the same order. Suppose one has a randomized block analysis of variance (RBF-.JK) with two levels of Factor A and four levels of Factor B. The cells would be entered in the order A1B1, A1B2, A1B3, A1B4, A2B1, A2B2, A2B3, A2B4. The blocks would be entered in the same order for each cell.

Observation summary tables for two randomized block factorial analyses of variance (RBF-.JK) are shown below:

Factor B

		B1	B2	B3	B4
		A1B1	A1B2	A1B3	A1B4
A1	Block 1	*1*	*3*	*5*	*7*
	Block 2	*2*	*4*	*6*	*8*
		A2B1	A2B2	A2B3	A2B4
A2	Block 1	*9*	*11*	*13*	*15*
	Block 2	*10*	*12*	*14*	*16*
		A3B1	A3B2	A3B3	A3B4
A3	Block 1	*17*	*19*	*21*	*23*
	Block 2	*18*	*20*	*22*	*24*

Factor B

		B1	B2	B3
		A1B1	A1B2	A1B3
	Block 1	*1*	*6*	*11*
	Block 2	*2*	*7*	*12*
A1	Block 3	*3*	*8*	*13*
	Block 4	*4*	*9*	*14*
	Block 5	*5*	*10*	*15*
		A2B1	A2B2	A2B3
	Block 1	*16*	*21*	*26*
	Block 2	*17*	*22*	*27*
A2	Block 3	*18*	*23*	*28*
	Block 4	*19*	*24*	*29*
	Block 5	*20*	*25*	*30*

The italicized numbers in the above examples show the order in which observations are entered in the randomized block factorial analysis of variance (RBF-.JK) programs.

Data for a randomized block factorial analysis of variance are given below. An actual analysis of variance would usually have more blocks.

Factor B

		B1	B2	B3	B4
A1	Block 1	131	91	52	51
	Block 2	115	107	65	100
A2	Block 1	80	72	76	45
	Block 2	112	83	83	95
A3	Block 1	19	35	46	91
	Block 2	40	15	65	50

When the program is run with the observations shown above, it should supply the following information:

Factor B

	B1	B2	B3	B4
A1	M11 = 123 V11 = 128	M12 = 99 V12 = 128	M13 = 58.5 V13 = 84.5	M14 = 75.5 V14 = 1200.5
A2	M21 = 96 V21 = 512	M22 = 77.5 V22 = 60.5	M23 = 79.5 V23 = 24.5	M24 = 70 V24 = 1250
A3	M31 = 29.5 V31 = 220.5	M32 = 25 V32 = 200	M33 = 55.5 V33 = 180.5	M34 = 70.5 V34 = 840.5

Note that Mjk is the mean for a cell, and Vjk is the variance for a cell with n - 1 degrees of freedom in the denominator of the variance formula.

The data for the analysis of variance could be entered in DATA statements as follows:

```
5000 DATA 2, 2, 3, 4
5010 DATA 131, 115, 91, 107, 52, 65, 51, 100
5020 DATA 80, 112, 72, 83, 76, 83, 45, 95
5030 DATA 19, 40, 35, 15, 46, 65, 91, 50
```

Means for levels of Factor A

A1 = 89
A2 = 80.75
A3 = 45.125

Means for levels of Factor B

B1 = 82.833333
B2 = 67.1666666
B3 = 64.499997
B4 = 71.9999998

Means for Blocks at levels of A

A1N1 = 81.25
A1N2 = 96.75
A2N1 = 68.25
A2N2 = 93.25
A3N1 = 47.75
A3N2 = 42.5

Means for Blocks at levels of B

B1N1 = 76.6666665
B1N2 = 88.9999999
B2N1 = 65.9999999
B2N2 = 68.3333332
B3N1 = 57.9999999
B3N2 = 70.9999998
B4N1 = 62.3333333
B4N2 = 81.6666665

Means for Blocks

Block 1 = 64.7499996

Block 2 = 77.4999994

	SS	DF	MS	F
TOTAL	21787.6242	23		
BLOCK	828.374964	1	828.374964	2.27739063
A	8699.24993	2	4349.62496	9.08775009
AXBLK	957.250126	2	478.625063	
B	1178.4583	3	392.819433	5.3053822
BXBLK	222.125052	3	74.041684	
AB	7080.41647	6	1180.06941	2.50922897
ABXBK	2821.74984	6	470.29164	
TOTAL	21787.6247	23		

Programs 6 and 12. Completely randomized factorial analysis of variance (requires equal n in all cells), CRF-JKL

When these programs are used for an independent groups three way analysis of variance (CRF-JKL), DATA statement 5000 should have five numbers. The first number is a 0 which indicates that there are no repeated measures (blocks) factors in the analysis of variance. The second number indicates the number of observations in a subgroup. The third, fourth, and fifth numbers indicate the number of levels of Factors A, B, and C, respectively. Consider the following example:

```
5000 DATA 0, 18, 6, 10, 12
```

The example indicates that Factor A has 6 levels, Factor B has 10 levels, and Factor C has 12 levels. All of the subgroups have 18 observations. Remember that the number of observations for each subgroup must be the same for these programs.

All of the observations for one subgroup must be entered before the observations for the next subgroup. Observations are entered with all of the subgroups at the first level of Factor A and the first level of Factor B entered first; i.e., A1B1C1, A1B1C2, A1B1C3 ... A1B1Cn. Then observations are entered at the first level of Factor A and the second level of Factor B; i.e., A1B2C1, A1B2C2, A1B2C3 ... A1B2Cn. This procedure is continued until all of the subgroups at level A1 are entered. Then the subgroups at level A2 are entered. First subgroups A2B1C1, A2B1C2, A2B1C3 ... A2B1Cn are entered. Then subgroups A2B2C1, A2B2C2, A2B2C3 ... A2B2Cn are entered.

The italicized numbers in the examples show the order in which observations from the cells would be entered in the analysis of variance programs. All of the observations for cell 1 in a given analysis would be entered before the observations in cell 2, etc. Observation summary tables for two independent groups three-way analyses of variance (CRF-JKL) are given below:

Factor C

		C1	C2
A1	B1	A1B1C1 *1*	A1B1C2 *2*
	B2	A1B2C1 *3*	A1B2C2 *4*
	B3	A1B3C1 *5*	A1B3C2 *6*
	B4	A1B4C1 *7*	A1B4C2 *8*
	B5	A1B5C1 *9*	A1B5C2 *10*
A2	B1	A2B1C1 *11*	A2B1C2 *12*
	B2	A2B2C1 *13*	A2B2C2 *14*
	B3	A2B3C1 *15*	A2B3C2 *16*
	B4	A2B4C1 *17*	A2B4C2 *18*
	B5	A2B5C1 *19*	A2B5C2 *20*
A3	B1	A3B1C1 *21*	A3B1C2 *22*
	B2	A3B2C1 *23*	A3B2C2 *24*
	B3	A3B3C1 *25*	A3B3C2 *26*
	B4	A3B4C1 *27*	A3B4C2 *28*
	B5	A3B5C1 *29*	A3B5C2 *30*

		C1	C2	C3	C4
A1	B1	A1B1C1 *1*	A1B1C2 *2*	A1B1C3 *3*	A1B1C4 *4*
	B2	A1B2C1 *5*	A1B2C2 *6*	A1B2C3 *7*	A1B2C4 *8*
	B3	A1B3C1 *9*	A1B3C2 *10*	A1B3C3 *11*	A1B3C4 *12*
A2	B1	A2B1C1 *13*	A2B1C2 *14*	A2B1C3 *15*	A2B1C4 *16*
	B2	A2B2C1 *17*	A2B2C2 *18*	A2B2C3 *19*	A2B2C4 *20*
	B3	A2B3C1 *21*	A2B3C2 *22*	A2B3C3 *23*	A2B3C4 *24*

Data for an independent groups three-way analysis of variance are given below. An independent groups three-way analysis of variance would usually have more observations per subgroup.

The data for the analysis of variance could be entered in data statements as follows:

```
5000 DATA 0, 2, 3, 4, 5
5010 DATA 5, 5, 9, 10, 12, 8, 8, 8, 9, 9
5020 DATA 6, 6, 8, 8, 11, 12, 13, 13, 10, 10
5030 DATA 11, 12, 9, 11, 9, 13, 13, 14, 11, 11
5040 DATA 12, 8, 14, 13, 14, 10, 14, 15, 12, 12
5050 DATA 10, 11, 11, 5, 10, 9, 11, 15, 13, 10
5060 DATA 9, 12, 12, 13, 8, 10, 14, 9, 10, 7
5070 DATA 12, 9, 12, 10, 10, 15, 14, 10, 13, 12
5080 DATA 12, 14, 12, 14, 11, 13, 14, 13, 13, 13
5090 DATA 15, 15, 14, 14, 13, 13, 15, 16, 15, 11
5100 DATA 16, 16, 18, 19, 17, 16, 16, 17, 16, 12
5110 DATA 20, 21, 16, 19, 19, 15, 18, 14, 17, 13
5120 DATA 18, 18, 17, 17, 20, 20, 19, 20, 18, 19
```

Factor C

		C1	C2	C3	C4	C5
	B1	5 5	9 10	12 8	8 8	9 9
	B2	6 6	8 8	11 12	13 13	10 10
A1	B3	11 12	9 11	9 13	13 14	11 11
	B4	12 8	14 13	14 10	14 15	12 12
	B1	10 11	11 5	10 9	11 15	13 10
	B2	9 12	12 13	8 10	14 9	10 7
A2	B3	12 9	12 10	10 15	14 10	13 12
	B4	12 14	12 14	11 13	14 13	13 13
	B1	15 15	14 14	13 13	15 16	15 11
	B2	16 16	18 19	17 16	16 17	16 12
A3	B3	20 21	16 19	19 15	18 14	17 13
	B4	18 18	17 17	20 20	19 20	18 19

		C1	C2	C3	C4	C5
	B1	M111= 5 V111= 0	M112= 9.5 V112= .5	M113= 10 V113= 8	M114= 8 V114= 0	M115= 9 V115= 0
	B2	M121= 6 V121= 0	M122= 8 V122= 0	M123= 11.5 V123= .5	M124= 13 V124= 0	M125= 10 V125= 0
A1	B3	M131= 11.5 V131= .5	M132= 10 V132= 2	M133= 11 V133= 8	M134=13.5 B134= .5	M135= 11 V135= 0
	B4	M141= 10 V141= 8	M142= 13.5 V142= .5	M143= 12 V143= 8	M144= 14.5 V144= .5	M145= 12 V145= 0
	B1	M211= 10.5 V211= .5	M212= 8 V212= 18	M213= 9.5 V213= .5	M214= 13 V214= 8	M215= 11.5 V215= 4.5
	B2	M221= 10.5 V221= 4.5	M222= 12.5 V222= .5	M223= 9 V223= 2	M224= 11.5 V224= 12.5	M225= 8.5 V225= 4.5
A2	B3	M231= 10.5 V231= 4.5	M232= 11 V232= 2	M233= 12.5 V233= 12.5	M234= 12 V234= 8	M235= 12.5 V235= .5
	B4	M241= 13 V241= 2	M242= 13 V242= 2	M243= 12 V243= 2	M244= 13.5 V244= .5	M245= 13 V245= 0
	B1	M311= 15 V311= 0	M312= 14 V312= 0	M313= 13 V313= 0	M314= 15.5 V314= .5	M315=13 V315= 8
	B2	M321= 16 V321= 0	M322= 18.5 V322= .5	M323= 16.5 V323= .5	M324= 16.5 V324= .5	M325= 14 V325= 8
A3	B3	M331= 20.5 V331= .5	M332= 17.5 V332= 4.5	M333= 17 V333= 8	M334= 16 V334= 8	M335= 15 V335= 8
	B4	M341= 18 V341= 0	M342= 17 V342= 0	M343= 20 V343= 0	M344= 19.5 V344= .5	M345= 18.5 V345= .5

Note that Mjkl is the mean for a cell, and Vjkl is the variance for a cell with n - 1 degrees of freedom in the denominator of the variance formula.

Means for AB interaction

	B1	B2	B3	B4
A1	8.3	9.7	11.4	12.4
A2	10.5	10.4	11.7	12.9
A3	14.1	16.3	17.2	18.6

Means for AC interaction

	C1	C2	C3	C4	C5
A1	8.125	10.25	11.125	12.25	10.5
A2	11.125	11.125	10.75	12.5	11.375
A3	17.375	16.75	16.625	16.875	15.125

Means for BC interaction

	C1	C2	C3	C4	C5
B1	10.1666666	10.4999999	10.8333333	12.1666666	11.1666666
B2	10.8333333	12.9999999	12.3333333	13.6666666	10.8333333
B3	14.1666666	12.8333333	13.4999999	13.8333332	12.8333332
B4	13.6666666	14.4999999	14.6666665	15.8333332	14.4999999

Means for levels of Factor A

A1 = 10.45

A2 = 11.375

A3 = 16.55

Means for levels of Factor B

B1 = 10.9666665

B2 = 12.1333331

B3 = 13.433333

B4 = 14.6333329

Means for levels of Factor C

C1 = 12.2083332

C2 = 12.7083331

C3 = 12.8333331

C4 = 13.8749997

C5 = 12.3333331

	SS	DF	MS	F
TOTAL	1553.79151	119		
A	864.61666	2	432.30833	148.644698
B	227.024963	3	75.6749876	26.020053
C	41.5833268	4	10.3958317	3.574498
AB	20.250951	6	3.3751585	1.16051295
AC	68.7166806	8	8.58958507	2.95343899
BC	34.6833682	12	2.89028068	0.993792785
ABC	122.415592	24	5.10064966	1.75380504
W.CEL	174.5	60	2.90833333	
TOTAL	1553.79152	119		

Programs 6 and 13. Split-plot analysis of variance (requires equal n in all cells), SPF-JK.L

When these programs are used for a split-plot factorial analysis of variance (SPF-JK.L) DATA statement 5000 should have five numbers. The first number is a 1 which indicates there is one repeated measures factor in the analysis of variance. The second number is the number of observations in each cell; i.e., at each ABC level. The third, fourth, and fifth numbers indicate the number of levels of Factors A, B, and C, respectively. Factor C (the third factor) is the repeated measures factor. Consider the following example:

```
5000 DATA 1, 4, 3, 5, 9
```

This example indicates that Factor A has 3 levels, Factor B has 5 levels and Factor C, the repeated measures factor, has 9 levels. All of the ABC cells have 4 observations (blocks within treatments). Remember that the number of observations for each cell must be the same for these programs.

All of the observations for one cell (level of A, B, and C) must be entered before the observations for the next cell. If observations in one cell are related to the observations in another cell (because of repeated measures Factor C), the related observations must always be entered in the same order. For example, if the first observation in a cell is from a given subject, this subject must provide the first observation in any cell in which he is represented. Observations are entered with all of the cells at the first level of Factor A and the first level of Factor B entered first; i.e., A1B1C1, A1B1C2, A1B1C3 ... A1B1Cn. Then observations are entered at the first level of Factor A and the second level of Factor B; i.e., A1B2C1, A1B2C2, A1B2C3 ... A1B2Cn. This procedure is continued until all of the cells at a given level of A1 are entered. Then the cells at level A2 are entered. First, cells A2B1C1, A2B1C2, A2B1C3 ... A2B1Cn are entered. Then cells A2B2C1, A2B2C2, A2B2C3 ... A2B2Cn are entered, etc. An observation summary table for a split-plot factorial analysis of variance (SPF-JK.L) is given below.

			Factor C			
			C1	C2	C3	C4
A1	B1		A1B1C1	A1B1C2	A1B1C3	A1B1C4
		Block 1	1	6	11	16
		Block 2	2	7	12	17
		Block 3	3	8	13	18
		Block 4	4	9	14	19
		Block 5	5	10	15	20
	B2		A1B2C1	A1B2C2	A1B2C3	A1B2C4
		Block 6	21	26	31	36
		Block 7	22	27	32	37
		Block 8	23	28	33	38
		Block 9	24	29	34	39
		Block 10	25	30	35	40
	B3		A1B3C1	A1B3C2	A1B3C3	A1B3C4
		Block 11	41	46	51	56
		Block 12	42	47	52	57
		Block 13	43	48	53	58
		Block 14	44	49	54	59
		Block 15	45	50	55	60
A2	B1		A2B1C1	A2B1C2	A2B1C3	A2B1C4
		Block 16	61	66	71	76
		Block 17	62	67	72	77
		Block 18	63	68	73	78
		Block 19	64	69	74	79
		Block 20	65	70	75	80
	B2		A2B2C1	A2B2C2	A2B2C3	A2B2C4
		Block 21	81	86	91	96
		Block 22	82	87	92	97
		Block 23	83	88	93	98
		Block 24	84	89	94	99
		Block 25	85	90	95	100
	B3		A2B3C1	A2B3C2	A2B3C3	A2B3C4
		Block 26	101	106	111	116
		Block 27	102	107	112	117
		Block 28	103	108	113	118
		Block 29	104	109	114	119
		Block 30	105	110	115	120

The italicized numbers in the table show the order in which observations from the cells would be entered in an analysis of variance program. All of the observations for cell 1 are entered before the observations in cell 2, etc.

Data for a split-plot factorial analysis of variance (SPF-JK.L) are given below. A split-plot analysis of variance (SPF-JK.L) would usually have more observations per cell. The data for the analysis of variance could be entered in DATA statements as follows:

```
5000 DATA 1, 3, 2, 4, 5
5010 DATA 5.1, 5.1, 5.3, 4.9, 5.5, 4.2
5020 DATA 10.1, 10.1, 10.3, 17.5, 16.4, 14.6, 20.9, 22.8, 20.9
5030 DATA 9.9, 9.2, 7.5, 10.5, 8.5, 9.5
5040 DATA 12.1, 13.1, 13.5, 17.1, 21.8, 20.0, 28.1, 24.8, 28.5
5050 DATA 15.9, 13.9, 19.5, 15.3, 13.9, 16.0
5060 DATA 21.4, 24.5, 17.9, 21.8, 26.7, 26.1, 31.4, 29.2, 32.7
5070 DATA 24.3, 18.1, 21.2, 18.0, 19.9, 17.8
5080 DATA 27.3, 28.6, 23.2, 30.5, 31.5, 31.5, 37.5, 34.0, 32.9
5090 DATA 11.0, 10.9, 9.6, 8.9, 9.4, 10.8
5100 DATA 20.1, 11.4, 12.9, 17.9, 23.3, 21.3, 22.7, 24.9, 23.4
5110 DATA 16.0, 14.6, 14.9, 14.7, 11.2, 16.1
5120 DATA 21.3, 19.0, 22.6, 22.6, 24.1, 28.4, 31.9, 31.0, 28.3
5130 DATA 19.7, 22.2, 17.2, 19.5, 19.2, 20.1
5140 DATA 23.8, 23.9, 24.7, 29.9, 31.7, 28.3, 37.6, 34.2, 35.9
5150 DATA 25.0, 23.2, 22.9, 28.7, 24.8, 25.4
5160 DATA 29.1, 28.6, 31.5, 33.2, 31.7, 34.0, 37.9, 40.0, 42.7
```

Factor C

			C1	C2	C3	C4	C5
A1	B1	Block 1	5.1	4.9	10.1	17.5	20.9
		Block 2	5.1	5.5	10.1	16.4	22.8
		Block 3	5.3	4.2	10.3	14.6	20.9
	B2	Block 4	9.9	10.5	12.1	17.1	28.1
		Block 5	9.2	8.5	13.1	21.8	24.8
		Block 6	7.5	9.5	13.5	20.0	28.5
	B3	Block 7	15.9	15.3	21.4	21.8	31.4
		Block 8	13.9	13.9	24.5	26.7	29.2
		Block 9	19.5	16.0	17.9	26.1	32.7
	B4	Block 10	24.3	18.0	27.3	30.5	37.5
		Block 11	18.1	19.9	28.6	31.5	34.0
		Block 12	21.2	17.8	23.2	31.5	32.9
A2	B1	Block 13	11.0	8.9	20.1	17.9	22.7
		Block 14	10.9	9.4	11.4	23.3	24.9
		Block 15	9.6	10.8	12.9	21.3	23.4
	B2	Block 16	16.0	14.7	21.3	22.6	31.9
		Block 17	14.6	11.2	19.0	24.1	31.0
		Block 18	14.9	16.1	22.6	28.4	28.3
	B3	Block 19	19.7	19.5	23.8	29.9	37.6
		Block 20	22.2	19.2	23.9	31.7	34.2
		Block 21	17.2	20.1	24.7	28.3	35.9
	B4	Block 22	25.0	28.7	29.1	33.2	37.9
		Block 23	23.2	24.8	28.6	31.7	40.0
		Block 24	22.9	25.4	31.5	34.0	42.7

When the program is run with the above data, it should supply the following information:

Factor C

		C1	C2	C3	C4	C5
A1	B1	5.16666666 .01333332	4.86666666 .423333331	10.1666666 .0133333332	16.1666665 2.14333331	21.2333332 1.20333332
	B2	8.86666666 1.5233331	9.49999999 .999999999	12.8999999 .519999999	19.6333332 5.62333331	27.1333333 4.12333331
	B3	16.4333333 8.05333331	15.0666666 1.14333331	21.2666665 10.9033333	24.8666666 7.1433333	31.0999999 3.12999997
	B4	21.1999999 9.60999997	18.5666666 1.34333332	26.3666666 7.94333329	31.1666661 .333333331	34.7999999 5.76999996
A2	B1	10.4999999 .609999999	9.69999999 .969999997	14.8 21.63	20.8333333 7.45333331	23.6666666 1.2633332
	B2	15.1666665 .543333333	13.9999999 6.36999997	20.9666666 3.3233333	25.0333332 9.0633333	30.3999999 3.50999997
	B3	19.6999999 6.24999997	19.6 .21	24.1333332 .243333333	29.9666665 2.89333331	35.8999999 2.88999997
	B4	23.6999999 1.28999999	26.2999999 4.40999996	29.7333333 2.40333331	32.9666665 1.36333331	40.1999999 5.78999996

Note that the top number in each cell is the cell mean. The bottom number in each cell is the variance for the cell with n - 1 degrees of freedom in the denominator of the variance formula.

Means for AB interaction

Factor B

	B1	B2	B3	B4
A1	11.5799999	15.6066665	21.7466665	26.4199997
A2	15.8999999	21.1133329	25.8599996	30.5799995

Means for AC interaction

Factor C

	C1	C2	C3	C4	C5
A1	12.9166665	11.9999999	17.6749999	22.9583332	28.6416666
A2	17.266665	17.399997	22.4083331	27.1999996	32.5416664

Means for BC interaction

Factor C

	C1	C2	C3	C4	C5
B1	7.83333332	7.28333332	12.4833333	18.4999999	22.5999999
B2	12.0166666	11.7499999	16.9333332	22.3333332	28.7666664
B3	18.0666665	17.3333333	22.6999998	27.4166665	33.4999997
B4	22.4333332	22.4333332	28.0499999	32.0666665	37.4999997

Means for levels of Factor A

A1 = 18.8383326

A2 = 23.3633321

Means for levels of Factor B

B1 = 13.7399998

B2 = 18.3599995

B3 = 23.8033327

B4 = 28.4999992

Means for levels of Factor C

C1 = 15.091663

C2 = 14.6999998

C3 = 20.0416662

C4 = 25.0791661

C5 = 30.591666

	SS	DF	MS	F
TOTAL	9153.90925	119		
BET	4376.98988	23		
A	614.26861	1	614.26861	242.383555
B	3712.35595	3	1237.45198	488.284776
AB	9.816797	3	3.27226566	1.291196
SWG.	40.5485337	16	2.53428335	
WITH	4776.91983	96		
C	4418.53509	4	1104.63377	263.467309
AC	7.86358	4	1.965895	0.468887589
BC	20.053619	12	1.67113491	0.398584064
ABC	62.1361275	12	5.17801062	1.2350125
CXSWG	268.331436	64	4.19267868	
TOTAL	9153.90972	119		

Programs 6 and 14. Split-plot analysis of variance (requires equal n in all cells), SPF-J.KL

When these programs are used for a split-plot factorial analysis of variance (SPF-J.KL) DATA statement 5000 should have five numbers. The first number is a 2 which indicates there are two repeated measures factors in the analysis of variance. The second number is the number of blocks represented at each ABC level. The third, fourth, and fifth numbers indicate the number of levels of Factors A, B, and C, respectively. Factors B and C (the second and third factors) are the repeated measures factors. Consider the following example:

```
5000 DATA 2, 6, 9, 4, 7
```

This example indicates that Factor A has 9 levels, Factor B (the first repeated measures factor) has 4 levels, and Factor C (the second repeated measures factor) has 7 levels. All the ABC cells have 6 observations (blocks within treatments). Remember that the number of observations for each cell must be the same for these programs.

All the observations for one cell (level of A, B, and C) must be entered before the observations in another cell. The related observations in another cell must always be entered in the same order. For example, if the first observation in a cell is from a given subject, this subject must provide the first observation in any cell in which she is represented.

Observations are entered with all of the cells at the first level of Factor A and the first level of Factor B entered first; i.e., A1B1C1, A1B1C2, A1B1C3 ... A1B1Cn. Then observations are entered at the first level of Factor A and the second level of Factor B; i.e., A1B2C1, A1B2C2, A1B2C3 ... A1B2Cn. This procedure is continued until all of the cells at a given level of Factor A are entered. Then the cells at the second level of Factor A are entered. First, cells A2B1C1, A2B1C2, A2B1C3 ... A2B1Cn are entered. Then cells A2B2C1, A2B2C2, A2B2C3 ... A2B2Cn are entered, etc. An observation summary table for a split-plot factorial analysis of variance (SPF-J.KL) is given below:

			Factor C			
			C1	C2	C3	C4
A1	B1		A1B1C1	A1B1C2	A1B1C3	A1B1C4
		Block 1	1	6	11	16
		Block 2	2	7	12	17
		Block 3	3	8	13	18
		Block 4	4	9	14	19
		Block 5	5	10	15	20
	B2		A1B2C1	A1B2C2	A1B2C3	A1B2C4
		Block 6	21	26	31	36
		Block 7	22	27	32	37
		Block 8	23	28	33	38
		Block 9	24	29	34	39
		Block 10	25	30	35	40
A2	B1		A2B1C1	A2B1C2	A2B1C3	A2B1C4
		Block 1	41	46	51	56
		Block 2	42	47	52	57
		Block 3	43	48	53	58
		Block 4	44	49	54	59
		Block 5	45	50	55	60
	B2		A2B2C1	A2B2C2	A2B2C3	A2B2C4
		Block 6	61	66	71	76
		Block 7	62	67	72	77
		Block 8	63	68	73	78
		Block 9	64	69	74	79
		Block 10	65	70	75	80
A3	B1		A3B1C1	A3B1C2	A3B1C3	A3B1C4
		Block 1	81	86	91	96
		Block 2	82	87	92	97
		Block 3	83	88	93	98
		Block 4	84	89	94	99
		Block 5	85	90	95	100
	B2		A3B2C1	A3B2C2	A3B2C3	A3B2C4
		Block 6	101	106	111	116
		Block 7	102	107	112	117
		Block 8	103	108	113	118
		Block 9	104	109	114	119
		Block 10	105	110	115	120

The italicized numbers in the table show the order in which observations from the cells would be entered in the analysis of variance programs. All the observations from cell 1 (A1B1C1) would be entered before the observations in cell 2, etc.

Data for a split-plot factorial analysis of variance (SPF-J.KL) are given below. A split-plot analysis of variance would usually have more observations per cell.

Factor C

			C1	C2	C3	C4	C5
A1	B1	Block 1	61	55	41	32	40
		Block 2	39	80	56	16	33
		Block 3	73	22	57	47	38
		Block 4	77	61	59	36	36
	B2	Block 5	90	79	84	111	59
		Block 6	58	87	83	100	111
		Block 7	47	88	58	73	67
		Block 8	76	88	74	85	61
A2	B1	Block 1	78	73	82	89	71
		Block 2	81	36	50	77	40
		Block 3	81	77	49	40	56
		Block 4	100	56	72	30	77
	B2	Block 5	33	46	16	29	51
		Block 6	75	86	48	38	70
		Block 7	45	67	52	42	65
		Block 8	64	19	36	44	9
A3	B1	Block 1	85	40	84	70	112
		Block 2	63	23	67	100	127
		Block 3	75	77	97	81	97
		Block 4	63	82	86	76	102
	B2	Block 5	18	42	52	23	67
		Block 6	21	47	57	34	52
		Block 7	21	34	44	44	40
		Block 8	40	22	32	57	30

The data for the analysis of variance could be entered in DATA statements as follows:

```
5000 DATA 2, 4, 3, 2, 5
5010 DATA 61, 39, 73, 77, 55, 80, 22, 61
5020 DATA 41, 56, 57, 59, 32, 16, 47, 36, 40, 33, 38, 36
5030 DATA 90, 58, 47, 76, 79, 87, 88, 88
5040 DATA 84, 83, 58, 74, 111, 100, 73, 85, 59, 111, 67, 61
5050 DATA 78, 81, 81, 100, 73, 36, 77, 56
5060 DATA 82, 50, 49, 72, 89, 77, 40, 30, 71, 40, 56, 77
5070 DATA 33, 75, 45, 64, 46, 86, 67, 19
5080 DATA 16, 48, 52, 36, 29, 38, 42, 44, 51, 70, 65, 9
5090 DATA 85, 63, 75, 63, 40, 23, 77, 82
5100 DATA 84, 67, 97, 86, 70, 100, 81, 76, 112, 127, 97, 102
5110 DATA 18, 21, 21, 40, 42, 47, 34, 22
5120 DATA 52, 57, 44, 32, 23, 34, 44, 57, 67, 52, 40, 30
```

When the program is run with the above data, it should supply the following information:

Factor C

		C1	C2	C3	C4	C5
A1	B1	62.5 291.666666	54.5 582.999999	53.25 68.2499999	32.75 164.916666	36.75 8.91666666
	B2	67.75 362.916666	85.5 18.9999999	74.75 144.916666	92.25 278.249999	74.5 603.666666
A2	B1	85 101.999999	60.5 349.666666	63.25 268.916666	59 808.666666	61 273.999999
	B2	54.25 354.249999	54.5 826.999999	38 261.333332	38.25 44.2499999	48.75 766.916666
A3	B1	71.5 112.999999	55.5 820.333332	83.5 153.666666	81.75 168.249999	109.5 174.999999
	B2	25 101.999999	36.25 118.916666	46.25 118.916666	39.5 209.666666	47.25 254.249999

Note that the top number in each cell is the cell mean. The bottom number in each cell is the variance for the cell with n - 1 degrees of freedom in the denominator of the variance formula.

Means for AB interaction

Factor B

	B1	B2
A1	47.95	78.95
A2	65.75	46.75
A3	80.35	38.85

Means for AC interaction

Factor C

	C1	C2	C3	C4	C5
A1	65.125	70	64	62.5	55.625
A2	69.625	57.5	50.625	48.625	54.875
A3	48.25	45.875	64.875	60.625	78.375

Means for BC interaction

Factor C

	C1	C2	C3	C4	C5
B1	72.9999998	56.8333328	66.6666663	57.833333	69.0833329
B2	48.9999998	58.7499996	52.9999996	56.6666662	56.8333333

Means for levels of Factor A

A1 = 63.45

A2 = 56.25

A3 = 59.6

Means for levels of Factor B

B1 = 64.6833315

B2 = 54.8499981

Means for levels of Factor C

C1 = 60.9999995

C2 = 57.7916657

C3 = 59.8333325

C4 = 57.2499992

C5 = 62.9583325

	SS	DF	MS	F
TOTAL	72511.4661	119		
BET.	2099.66664	11		
A	1038.46666	2	519.23333	4.40359978
SWG.	1061.19998	9	117.911108	
WITH	70411.7992	108		
B	2900.83336	1	2900.83336	4.84505843
AB	27541.6863	2	13770.8431	23.000473
BXSWG	5388.47995	9	598.719994	
C	526.716688	4	131.679172	0.60821406
AC	8104.03305	8	1013.00413	4.67897346
CXSWG	7794.049	36	216.501361	
BC	2606.41648	4	651.66412	1.92185715
ABC	3343.81373	8	417.976716	1.23279076
BCXSG	12205.771	36	339.049194	
TOTAL	72511.4661	119		

Programs 6 and 15. Randomized block factorial analysis of variance (requires equal n in all cells), RBF-.JKL

When these programs are used for a randomized block factorial analysis of variance (RBF-.JKL), DATA statement 5000 should have five numbers. The first number is a 3 which indicates all three factors, A, B, and C, are repeated measures (block) factors. The second number is the number of blocks represented at each ABC level. The third, fourth, and fifth numbers indicate the number of levels of Factors A, B, and C, respectively. Consider the following example:

```
5000 DATA 3, 4, 7, 10, 12
```

The example indicates that Factor A has 7 levels, Factor B has 10 levels, and Factor C has 12 levels. All the ABC cells have 4 observations (blocks). Remember that the number of observations from each cell must be the same for these programs.

All the observations for one cell (level of A, B, and C) must be entered before the observations in another cell. The related observations must be entered in the same order for each cell. For example, if the first observation in a cell is from a given subject, this subject (block) must provide the first observation in all the cells.

Observations are entered with all the cells at the first level of Factor A and the first level of Factor B entered first; i.e., A1B1C1, A1B1C2, A1B1C3, ... A1B1Cn. Then observations are entered at the first level of Factor A and the second level of Factor B; i.e., A1B2C1, A1B2C2, A1B2C3 ... A1B2Cn. This procedure is continued until all the cells at a given level of A1 are entered. Then the cells at level A2 are entered. First cells A2B1C1, A2B1C2, A2B1C3 ... A2B1Cn are entered. Then cells A2B2C1, A2B2C2, A2B2C3 ... A2B2Cn are entered. An observation summary table for a randomized block factorial analysis of variance (RBF-.JKL) is given below:

			Factor C				
			C1	C2	C3	C4	C5
A1	B1		A1B1C1	A1B1C2	A1B1C3	A1B1C4	A1B1C5
		Block 1	1	3	5	7	9
		Block 2	2	4	6	8	10
	B2		A1B2C1	A1B2C2	A1B2C3	A1B2C4	A1B2C5
		Block 1	11	13	15	17	19
		Block 2	12	14	16	18	20
	B3		A1B3C1	A1B3C2	A1B3C3	A1B3C4	A1B3C5
		Block 1	21	23	25	27	29
		Block 2	22	24	26	28	30
	B4		A1B4C1	A1B4C2	A1B4C3	A1B4C4	A1B4C5
		Block 1	31	33	35	37	39
		Block 2	32	34	36	38	40
A2	B1		A2B1C1	A2B1C2	A2B1C3	A2B1C4	A2B1C5
		Block 1	41	43	45	47	49
		Block 2	42	44	46	48	50
	B2		A2B2C1	A2B2C2	A2B2C3	A2B2C4	A2B2C5
		Block 1	51	53	55	57	59
		Block 2	52	54	56	58	60
	B3		A2B3C1	A2B3C2	A2B3C3	A2B3C4	A2B3C5
		Block 1	61	63	65	67	69
		Block 2	62	64	66	68	70
	B4		A2B4C1	A2B4C2	A2B4C3	A2B4C4	A2B4C5
		Block 1	71	73	75	77	79
		Block 2	72	74	76	78	80
A3	B1		A3B1C1	A3B1C2	A3B1C3	A3B1C4	A3B1C5
		Block 1	81	83	85	87	89
		Block 2	82	84	86	88	90
	B2		A3B2C1	A3B2C2	A3B2C3	A3B2C4	A3B2C5
		Block 1	91	93	95	97	99
		Block 2	92	94	96	98	100
	B3		A3B3C1	A3B3C2	A3B3C3	A3B3C4	A3B3C5
		Block 1	101	103	105	107	109
		Block 2	102	104	106	108	110
	B4		A3B4C1	A3B4C2	A3B4C3	A3B4C4	A3B4C5
		Block 1	111	113	115	117	119
		Block 2	112	114	116	118	120

The italicized numbers in the table show the order in which observations from the cells would be entered in the analysis of variance programs. All the observations from cell 1 (A1B1C1) would be entered before the observations from cell 2, etc.

Data for a randomized block factorial analysis of variance (RBF-.JKL) are given below. A randomized block factorial analysis of variance (RBF-.JKL) would usually have more blocks.

The data for the analysis of variance could be entered in DATA statements as follows:

```
5000 DATA 3, 3, 2, 4, 5
5010 DATA 6, 4, 6, 5, 6, 6, 3, 4, 3
5020 DATA 3, 6, 7, 2, 3, 6
5030 DATA 4, 9, 10, 2, 6, 5, 5, 10, 3
5040 DATA 10, 3, 2, 5, 8, 7
5050 DATA 8, 9, 10, 3, 5, 9, 7, 9, 9,
5060 DATA 8, 7, 7, 7, 3, 10
5070 DATA 5, 8, 8, 8, 7, 10, 9, 9, 10
5080 DATA 7, 6, 4, 10, 7, 12
5090 DATA 4, 5, 2, 4, 6, 4, 0, 5, 6
5100 DATA 5, 5, 11, 4, 5, 7
5110 DATA 1, 7, 9, 5, 8, 10, 7, 10, 4
5120 DATA 6, 4, 5, 4, 6, 5
5130 DATA 6, 7, 8, 8, 7, 4, 7, 10, 10
5140 DATA 9, 4, 6, 5, 6, 7
5150 DATA 8, 11, 12, 10, 7, 11, 3, 6, 6
5160 DATA 7, 7, 8, 4, 8, 6
```

Factor C

			C1	C2	C3	C4	C5
A1	B1	Block 1	6	5	3	3	2
		Block 2	4	6	4	6	3
		Block 3	6	6	3	7	6
	B2	Block 1	4	2	5	10	5
		Block 2	9	6	10	3	8
		Block 3	10	5	3	2	7
	B3	Block 1	8	3	7	8	7
		Block 2	9	5	9	7	3
		Block 3	10	9	9	7	10
	B4	Block 1	5	8	9	7	10
		Block 2	8	7	9	6	7
		Block 3	8	10	10	4	12
A2	B1	Block 1	4	4	0	5	4
		Block 2	5	6	5	5	5
		Block 3	2	4	6	11	7
	B2	Block 1	1	5	7	6	4
		Block 2	7	8	10	4	6
		Block 3	9	10	4	5	5
	B3	Block 1	6	8	7	9	5
		Block 2	7	7	10	4	6
		Block 3	8	4	10	6	7
	B4	Block 1	8	10	3	7	4
		Block 2	11	7	6	7	8
		Block 3	12	11	6	8	6

When the program is run with the above data, it should supply the following information:

Factor C

		C1	C2	C3	C4	C5
A1	B1	5.3333333 1.33333332	5.66666666 .333333333	3.33333333 .333333332	5.33333333 4.3333333	3.66666666 4.3333333
	B2	7.66666666 10.3333332	4.33333332 4.3333333	5.99999999 12.9999998	4.99999999 18.9999998	6.66666665 2.3333333
	B3	8.99999999 .999999999	5.66666666 9.33333328	8.33333333 1.33333332	7.33333332 .333333333	6.66666666 12.3333332
	B4	6.99999998 2.99999997	8.33333332 2.33333333	9.33333333 .333333332	5.66666666 2.3333333	9.66666666 6.33333333
A2	B1	3.66666665 2.3333333	4.66666666 1.33333331	3.66666666 10.3333332	6.99999998 11.9999999	5.33333332 2.33333333
	B2	5.66666665 17.3333331	7.66666665 6.3333333	6.99999999 8.99999996	4.99999999 .999999999	4.99999999 .999999999
	B3	6.99999999 .999999999	6.3333332 4.3333333	8.99999999 2.99999996	6.33333333 6.33333331	5.99999999 .999999999
	B4	10.3333333 4.33333331	9.33333332 4.3333333	5 3	7.3333332 .333333333	5.99999999 3.99999996

Note that the top number in each cell is the cell mean. The bottom number in each cell is the variance for the cell with n - 1 degrees of freedom in the denominator of the variance formula.

Means for AB interaction

Factor B

	B1	B2	B3	B4
A1	4.66666664	5.93333329	7.39999995	7.99999994
A2	4.86666662	6.06666661	6.93333327	7.59999995

Means for AC interaction

Factor C

	C1	C2	C3	C4	C5
A1	7.24999996	5.99999996	6.74999998	5.8333333	6.6666664
A2	6.66666664	6.99999995	6.16666664	6.41666663	5.5833333

Means for BC interaction

Factor C

	C1	C2	C3	C4	C5
B1	4.49999998	5.16666665	3.49999999	6.16666665	4.49999998
B2	6.66666664	5.99999998	6.49999997	4.99999998	5.83333331
B3	7.99999998	5.99999998	8.66666664	6.83333331	6.33333331
B4	8.66666665	8.8333333	7.16666666	6.49999997	7.83333331

Means for levels of Factor A

A1 = 6.49999983

A2 = 6.36666647

Means for levels of Factor B

B1 = 4.76666658

B2 = 5.9999999

B3 = 7.16666659

B4 = 7.79999992

Means for levels of Factor C

C1 = 6.95833326

C2 = 6.49999991

C3 = 6.45833327

C4 = 6.12499992

C5 = 6.12499992

Means for Blocks

Block 1 = 5.6

Block 2 = 6.57

Block 3 = 7.125

	SS	DF	MS	F
TOTAL	757.466615	119		
BLOCK	47.7166664	2	23.8583332	5.5011585
A	0.53333354	1	0.53333354	0.378698544
AXBLK	2.8166654	2	1.4083327	
B	161.133332	3	53.7111106	12.9858962
BXBLK	24.816667	6	4.13611116	
C	11.3	4	2.825	0.340105356
CXBLK	66.4499973	8	8.30624966	
AB	2.73338115	3	0.91112705	0.33920039
ABXBK	16.1166156	6	2.6861026	
AC	18.6333324	4	4.6583331	1.11188475
ACXBK	33.5166613	8	4.18958266	
BC	76.3666651	12	6.36388875	1.08565354
BCXBK	140.683306	24	5.86180441	
ABC	100.766582	12	8.39721516	3.74017318
ABCXB	53.8833776	24	2.24514073	
TOTAL	757.466584	119		

Program 16. Latin square design (requires equal n in all cells), LS-K

The Latin square design (LS-K) has two "factors" that are nuisance variables (A and B) and one factor that is of experimental interest (Factor C). Factors A, B, and C must have the same number of levels. If Factor C has 5 levels, Factors A, and B must each have 5 levels. The following table shows the cells for a possible Latin square design with 4 levels of each factor.

	B1	B2	B3	B4
A1	C1	C2	C3	C4
A2	C2	C3	C4	C1
A3	C3	C4	C1	C2
A4	C4	C1	C2	C3

Factor A is represented by the rows of the table and Factor B is represented by the columns of the table. Each level of Factor C occurs one time in each row. Each level of Factor C also occurs one time in each column. One could select another arrangement for Factor C that would have each level of Factor C represented one time in each row and one time in each column. The following table shows another arrangement for Factor C.

	B1	B2	B3	B4
A1	C2	C3	C1	C4
A2	C4	C2	C3	C1
A3	C3	C1	C4	C2
A4	C1	C4	C2	C3

Texts such as Kirk (1982) show how to randomly select the arrangement for Factor C in a Latin square experiment. Each of the cells in the above examples can have two or more observations. These observations should all be independent of each other. The following table shows two observations in each cell for a Latin square design (LS-K).

	B1	B2	B3	B4
A1	C3 6 6	C2 7 5	C1 3 5	C4 10 10
A2	C4 6 10	C3 9 9	C2 5 7	C1 7 7
A3	C2 6 8	C1 8 8	C4 9 13	C3 10 12
A4	C1 7 9	C4 13 15	C3 11 11	C2 10 12

Data statement 5000 should have two numbers, the number of observations per cell, and the number of levels of the experimental treatment (this number is also the number of levels for either nuisance variable). Note that the present program requires each cell have the same number of observations.

Data statement 5000 for the above example should be:

```
5000 DATA 2, 4
```

The observations for the cells are entered in the program row by row. Thus, the observations in the first row (level A1) are entered before the observations in the second row (level A2). Within a row the observations are entered with B1 observations before B2 observations, etc. Before the observations for a cell are entered, the level of C for the cell must be entered. In the present example DATA statement 5010 would be:

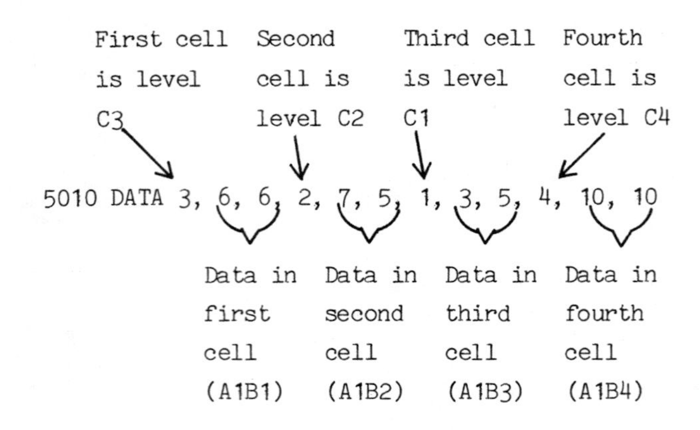

All of the DATA statements in the present example could be entered as follows:

```
5000 DATA 2, 4
5010 DATA 3, 6, 6, 2, 7, 5, 1, 3, 5, 4, 10, 10
5020 DATA 4, 6, 10, 3, 9, 9, 2, 5, 7, 1, 7, 7
5030 DATA 2, 6, 8, 1, 8, 8, 4, 9, 13, 3, 10, 12
5040 DATA 1, 7, 9, 4, 13, 15, 3, 11, 11, 2, 10, 12
```

When the Latin square program is run with the above data, it will supply the following information:

Factor B

	B1	B2	B3	B4
A1	M113 = 6 V113 = 0	M122 = 6 V122 = 2	M131 = 4 V131 = 2	M144 = 10 V144 = 0
A2	M214 = 8 V214 = 8	M223 = 9 V223 = 0	M232 = 6 V232 = 2	M241 = 7 V241 = 0
A3	M312 = 7 V312 = 2	M321 = 8 V321 = 0	M334 = 11 V334 = 8	M343 = 11 V343 = 2
A4	M411 = 8 V411 = 2	M424 = 14 V424 = 2	M433 = 11 V433 = 0	M442 = 11 V442 = 2

Note that mjkl is the mean for a cell, and Vjkl is the variance for a cell with n - 1 degrees of freedom in the denominator of the variance formula.

Means for levels of first nuisance variable (Factor A)

A1 = 6.5
A2 = 7.5
A3 = 9.25
A4 = 11

Means for levels of second nuisance variable (Factor B)

B1 = 7.25
B2 = 9.25
B3 = 8
B4 = 9.75

Means for treatment of interest (Factor C)

C1 = 6.75
C2 = 7.5
C3 = 9.25
C4 = 10.75

	SS	DF	MS	F
TOTAL	235.874994	31		
BETA	94.375	3	31.4583333	15.7291666
BETB	31.375	3	10.4583333	5.22916665
BETC	77.375	3	25.7916666	12.8958333
RES.	0.749994	6	0.124999	0.0624995
W.CEL	32	16	2	

Program 17. Analysis of covariance--one factor (requires equal n in all cells)

The present analysis of covariance program is designed to provide information that is similar to the information provided by an independent groups one-way analysis of variance. Interpretation of comparisons among means is modified by pretest observations for each of the subjects. The program requires a pretest observation (x) and a treatment observation (y) for each of the subjects. The number of pairs of observations for each treatment level must be the same. Consider the following example:

Treatment level 1			Treatment level 2			Treatment level 3		
	x	y		x	y		x	y
S1	90	70	S6	97	90	S11	130	120
S2	80	65	S7	85	80	S12	122	118
S3	77	72	S8	60	60	S13	150	141
S4	48	40	S9	112	110	S14	141	130
S5	35	29	S10	130	120	S15	127	125

DATA statement 5000 should have two numbers, the number of pairs of observations for a treatment level and the number of treatment levels. Thus, DATA statement 5000 for the present example would be:

```
5000 DATA 5, 3
```

The observations are entered one pair at a time. The x observation for a pair is entered before the y observation. All of the pairs for one treatment level are entered before the pairs for the next treatment level. The DATA statements for the example would be:

```
5000 DATA 5, 3
5010 DATA 90, 70, 80, 65, 77, 72, 48, 40, 35, 29
5020 DATA 97, 90, 85, 80, 60, 60, 112, 110, 130, 120
5030 DATA 130, 120, 122, 118, 150, 141, 141, 130, 127, 125
```

When the analysis of covariance is completed, two F-ratios are presented. The degrees of freedom for the numerator are presented first. The first F-ratio indicates if the regression lines for the observations are parallel. If this F-ratio is statistically significant, the second F-ratio comparing treatment levels for variable y may be meaningless. The program also supplies means for the x observations at each treatment level, and means for the y observations at each treatment level. The observations used in the example would provide the following F-ratios and means:

```
Mean treatment X1 = 66
Mean treatment Y1 = 55.2
SS within X1 = 2178
SS within Y1 = 1514.8
SS within X1Y1 = 1763

Mean treatment X2 = 96.8
Mean treatment Y2 = 92
SS within X2 = 2826.8
SS within Y2 = 2280
SS within X2Y2 = 2522

Mean treatment X3 = 134
Mean treatment Y3 = 126.8
SS within X3 = 514
SS within Y3 = 338.8
SS within X3Y3 = 395
```

F-test of differences between group regression coefficients; F = 0.353198919 with 2 (numerator) and 9 (denominator) degrees of freedom. If the above F-ratio is statistically significant at the significance level that was chosen, do not continue with the analysis of covariance. (The F-ratio is not statistically

significant in the present example.)
F-test for analysis of covariance; F = 7.14979097 with 2 (numerator) and 11 (denominator) degrees of freedom.

Chapter 4

Analysis of Variance Programs

Program 1. Completely randomized analysis of variance (does not require equal n in all cells), CR-K.

```
0010 REM THIS PROGRAM IS FOR AN INDEPENDENT GROUPS ONE-WAY ANALYSIS
0020 REM OF VARIANCE (COMPLETELY RANDOMIZED ANALYSIS OF VARIANCE,
0030 REM CR-K) WITH (OR WITHOUT) EQUAL N PER GROUP
0040 REM SEE APPROPRIATE EXAMPLE FOR ENTERING DATA
0050 PRINT
0060 READ T
0070 REM CALCULATE NUMBER OF DATA POINTS (N) AND
0080 REM NUMBER IN EACH GROUP (N1)
0090 LET N = 0
0100 LET N1 = 0
0110 FOR J = 1 TO T
0120 READ Q
0130 IF Q > 2E22 THEN 170
0140 LET N = N + 1
0150 LET N1 = N1 + 1
0160 GOTO 120
0170 PRINT "N";J;" = ";N1
0180 GOSUB 3060
0190 LET N1 = 0
0200 NEXT J
0210 REM CALCULATE GRAND MEAN
0220 RESTORE
0230 READ Q
0240 FOR I = 1 TO N + T - 1
0250 READ Q
0260 IF Q > 2E22 THEN 280
0270 LET M = M + Q/N
0280 NEXT I
0290 REM CALCULATE MEAN FOR EACH GROUP (M2), VAR. FOR EACH
0300 REM GROUP (V3), SSTOTAL (S1), SSBET. (S2), AND SSWITH. CELL (S3)
0310 LET S1 = 0
0320 LET S2 = 0
0330 LET S3 = 0
0340 LET V3 = 0
0350 LET M2 = 0
0360 LET N1 = 0
0370 LET L = 0
0380 FOR J = 1 TO T
0390 RESTORE
0400 READ Q
0410 IF L = 0 THEN 430
0420 GOSUB 3010
0430 READ Q
0440 IF Q > 2E22 THEN 470
0450 LET N1 = N1 + 1
0460 GOTO 430
0470 RESTORE
0480 READ Q
0490 IF L = 0 THEN 510
0500 GOSUB 3010
0510 READ Q
0520 IF Q > 2E22 THEN 550
0530 LET M2 = M2 + Q/N1
0540 GOTO 510
0550 RESTORE
0560 READ Q
```

Program 1. (cont.)

```
0570 IF L = 0 THEN 590
0580 GOSUB 3010
0590 READ Q
0600 IF Q > 2E22 THEN 650
0610 LET S1 = S1 + (Q - M)*(Q - M)
0620 LET S3 = S3 + (Q - M2)*(Q - M2)
0630 LET V3 = V3 + (Q - M2)*(Q - M2)/(N1 - 1)
0640 GOTO 590
0650 PRINT "MEAN GROUP ";J;" = ";M2;
0660 PRINT "VAR. GROUP ";J;" = ";V3
0670 GOSUB 3050
0680 LET S2 = S2 + (M2 - M)*(M2 - M)*N1
0690 LET M2 = 0
0700 LET V3 = 0
0710 LET L = L + N1 + 1
0720 LET N1 = 0
0730 NEXT J
0740 LET T1 = 13
0750 GOSUB 3070
0760 PRINT TAB(9);"SS";TAB(22);"DF";TAB(35);"MS";TAB(48);"F"
0770 PRINT "TOTAL";TAB(9);S1;TAB(22);N - 1
0780 PRINT " BET.";TAB(9);S2;TAB(22);T - 1;TAB(35);S2/(T - 1);
0790 PRINT TAB(48);(S2/(T - 1))/(S3/(N - T))
0800 PRINT " W.CEL";TAB(9);S3;TAB(22);N - T;TAB(35);S3/(N - T)
0810 PRINT "TOTAL";TAB(9);S2 + S3;TAB(22);N - 1
0820 GOTO 9999
3000 REM SUBROUTINE TO ADVANCE DATA COUNTER
3010 FOR I = 1 TO L
3020 READ Q
3030 NEXT I
3040 RETURN
3050 REM SUBROUTINE TO STOP EXECUTION UNTIL DATA ARE TRANSCRIBED
3060 LET T1 = T1 + 1
3070 IF T1 < 10 THEN RETURN
3080 PRINT
3090 PRINT "ENTER 1 FOLLOWED BY CARRIAGE RETURN TO CONTINUE."
3100 INPUT Q
3110 LET T1 = 0
3120 RETURN
9999 END
```

Program 2. Completely randomized factorial analysis of variance (does not require equal n in all cells), CRF-JK.

```
0010 REM THIS PROGRAM IS FOR AN INDEPENDENT GROUPS TWO-WAY
0020 REM ANALYSIS OF VARIANCE (COMPLETELY RANDOMIZED
0030 REM FACTORIAL ANALYSIS OF VARIANCE, CRF-JK) WITH (OR
0040 REM WITHOUT) EQUAL N PER GROUP
0050 REM SEE APPROPRIATE EXAMPLE FOR ENTERING DATA
0060 PRINT
0070 READ P,Q
0080 REM CALCULATE NUMBER OF DATA POINTS (N), NUMBER IN
0090 REM EACH GROUP (N1), UNWEIGHTED GRAND MEAN (M),
0100 REM AND SUM OF THE RECIPROCALS OF NUMBER OF SCORES
0110 REM IN EACH GROUP (N2)
0120 LET M = 0
0130 LET N = 0
0140 LET N1 = 0
0150 LET N2 = 0
0160 LET L = 0
0170 FOR J = 1 TO P
0180 FOR K = 1 TO Q
0190 READ R
0200 IF R > 2E22 THEN 240
0210 LET N = N + 1
0220 LET N1 = N1 + 1
0230 GOTO 190
0240 GOSUB 3010
0250 IF L = 0 THEN 270
0260 GOSUB 3050
0270 READ R
0280 IF R > 2E22 THEN 310
0290 LET M = M + R/N1/P/Q
0300 GOTO 270
0310 LET N2 = N2 + 1/N1
0320 LET L = L + N1 + 1
0330 PRINT "N FOR A ";J;" B ";K;" = ";N1
0340 GOSUB 3110
0350 LET N1 = 0
0360 GOSUB 3010
0370 GOSUB 3050
0380 NEXT K
0390 NEXT J
0400 REM CALCULATE MEANS FOR EACH CELL (M2), VAR. FOR EACH CELL
0410 REM (V5), SSWITHIN CELLS (S5), MEANS FOR LEVELS OF FACTOR
0420 REM A(M1), SSA (S2), AND INTERMEDIATE VALUE FOR SSAB (S4)
0430 LET M1 = 0
0440 LET M2 = 0
0450 LET S2 = 0
0460 LET S4 = 0
0470 LET S5 = 0
0480 LET V5 = 0
0490 LET N1 = 0
0500 LET L = 0
0510 GOSUB 3010
0520 FOR J = 1 TO P
0530 FOR K = 1 TO Q
0540 READ R
0550 IF R > 2E22 THEN 580
0560 LET N1 = N1 + 1
0570 GOTO 540
0580 GOSUB 3010
0590 IF L = 0 THEN 610
0600 GOSUB 3050
0610 READ R
0620 IF R > 2E22 THEN 650
0630 LET M2 = M2 + R/N1
0640 GOTO 610
0650 LET S4 = S4 + (M2 - M)*(M2 - M)
0660 LET M1 = M1 + M2/Q
0670 PRINT "MEAN A ";J;" B ";K;" = ";M2;
0680 GOSUB 3110
0690 GOSUB 3010
0700 IF L = 0 THEN 720
```

Program 2. (cont.)

```
0710 GOSUB 3050
0720 READ R
0730 IF R > 2E22 THEN 770
0740 LET S5 = S5 + (R - M2)*(R - M2)
0750 LET V5 = V5 + (R - M2)*(R - M2)/(N1 - 1)
0760 GOTO 720
0770 LET M2 = 0
0780 PRINT "VAR. A ";J;" B ";K;" = ";V5
0790 GOSUB 3110
0800 LET V5 = 0
0810 LET L = L + N1 + 1
0820 LET N1 = 0
0830 GOSUB 3010
0840 GOSUB 3050
0850 NEXT K
0860 PRINT "MEAN LEVEL A";J;" = ";M1
0870 GOSUB 3110
0880 LET S2 = S2 + (M1 - M)*(M1 - M)*Q
0890 LET M1 = 0
0900 NEXT J
0910 REM CALCULATE MEANS FOR LEVELS OF FACTOR B(M3),
0920 REM SSB (S3), AND COMPLETED VALUE FOR SSAB (S4)
0930 LET H = 0
0940 LET L = 0
0950 LET M3 = 0
0960 LET M5 = 0
0970 LET N1 = 0
0980 LET S3 = 0
0990 FOR K = 1 TO Q
1000 FOR J = 1 TO P
1010 GOSUB 3010
1020 IF H = 0 THEN 1050
1030 LET G = H
1040 GOSUB 3180
1050 IF L = 0 THEN 1080
1060 LET G = L
1070 GOSUB 3180
1080 READ R
1090 IF R > 2E22 THEN 1120
1100 LET N1 = N1 + 1
1110 GOTO 1080
1120 GOSUB 3010
1130 IF H = 0 THEN 1160
1140 LET G = H
1150 GOSUB 3180
1160 IF L = 0 THEN 1190
1170 LET G = L
1180 GOSUB 3180
1190 READ R
1200 IF R > 2E22 THEN 1230
1210 LET M5 = M5 + R/N1
1220 GOTO 1190
1230 LET M3 = M3 + M5/P
1240 LET N1 = 0
1250 LET M5 = 0
1260 LET L = L + Q
1270 NEXT J
1280 PRINT "MEAN LEVEL B ";K;" = ";M3
1290 LET S3 = S3 + (M3 - M)*(M3 - M)*P
1300 LET M3 = 0
1310 LET H = H + 1
1320 LET L = 0
1330 NEXT K
1340 LET S4 = S4 - S2 - S3
1350 LET S5 = S5*N2/P/Q/(N - P*Q)
1360 PRINT TAB(9);"SS";TAB(22);"DF";TAB(35);"MS";TAB(48);"F"
1370 PRINT "A";TAB(9);S2;TAB(22);P - 1;TAB(35);S2/(P - 1);
1380 PRINT TAB(48);S2/(P - 1)/S5
1390 PRINT "B";TAB(9);S3;TAB(22);Q - 1;TAB(35);S3/(Q - 1);
1400 PRINT TAB(48);S3/(Q - 1)/S5
1410 PRINT "AB";TAB(9);S4;TAB(22);(P - 1)*(Q - 1);TAB(35);
1420 PRINT S4/(P - 1)/(Q - 1);TAB(48);S4/(P - 1)/(Q - 1)/S5
```

Program 2. (cont.)

```
1430 PRINT "WITH";TAB(22);N - P*Q;TAB(35);S5
1440 GOTO 9999
3000 REM RESTORE DATA POINTER AND ADVANCE DATA POINTER
3010 RESTORE
3020 READ R,R
3030 RETURN
3040 REM ADVANCE DATA POINTER
3050 FOR L1 = 1 TO L
3060 READ R
3070 IF R > 3E24 THEN 3090
3080 NEXT L1
3090 RETURN
3100 REM SUBROUTINE TO STOP EXECUTION UNTIL DATA ARE TRANSCRIBED
3110 LET T = T + 1
3120 IF T <= 12 THEN RETURN
3130 PRINT "ENTER 1 FOLLOWED BY CARRIAGE RETURN TO CONTINUE."
3140 INPUT R1
3150 LET T = 0
3160 RETURN
3170 REM ADVANCE POINTER ONE OR MORE SUBGROUPS
3180 FOR L2 = 1 TO G
3190 READ R
3200 IF R > 2E22 THEN 3220
3210 GOTO 3190
3220 NEXT L2
3230 RETURN
9999 END
```

Program 3. Completely randomized factorial analysis of variance (does not require equal n in all cells), CRF-JKL.

```
0010 REM THIS PROGRAM IS FOR AN INDEPENDENT GROUPS THREE-WAY
0020 REM ANALYSIS OF VARIANCE (COMPLETELY RANDOMIZED
0030 REM FACTORIAL ANALYSIS OF VARIANCE, CRF-JKL) WITH (OR
0040 REM WITHOUT) EQUAL N PER GROUP
0050 REM SEE APPROPRIATE EXAMPLE FOR ENTERING DATA
0060 PRINT
0070 READ P,Q,R
0080 REM CALCULATE NUMBER OF DATA POINTS (N), NUMBER IN EACH
0090 REM GROUP (N1), UNWEIGHTED GRAND MEAN (M), AND SUM OF THE
0100 REM RECIPROCALS OF NUMBER OF SCORES IN EACH GROUP (N2)
0110 LET L1 = 0
0120 LET M = 0
0130 LET N = 0
0140 LET N1 = 0
0150 LET N2 = 0
0160 LET T = 0
0170 FOR J = 1 TO P
0180 FOR K = 1 TO Q
0190 FOR L = 1 TO R
0200 READ S
0210 IF S > 2E22 THEN 250
0220 LET N = N + 1
0230 LET N1 = N1 + 1
0240 GOTO 200
0250 GOSUB 3010
0260 IF L1 = 0 THEN 280
0270 GOSUB 3050
0280 READ S
0290 IF S > 2E22 THEN 320
0300 LET M = M + S/N1/P/Q/R
0310 GOTO 280
0320 LET N2 = N2 + 1/N1
0330 LET L1 = L1 + N1 + 1
0340 PRINT "N FOR A";J;"B";K;"C";L;" = ";N1
0350 GOSUB 3110
0360 LET N1 = 0
0370 GOSUB 3010
0380 GOSUB 3050
0390 NEXT L
0400 NEXT K
0410 NEXT J
0420 REM CALCULATE MEAN EACH LEVEL FACTOR A (M2), SSA(S2),
0430 REM MEAN EACH LEVEL AB INTERACTION (M5), INTERMEDIATE
0440 REM VALUE FOR SSAB (S5), MEAN EACH LEVEL ABC INTERACTION
0450 REM (M8), INTERMEDIATE VALUE FOR SSABC (S8), AND
0460 REM SSWITHIN CELL (S9)
0470 LET G = 0
0480 LET H = 0
0490 LET M2 = 0
0500 LET M5 = 0
0510 LET M8 = 0
0520 LET N1 = 0
0530 LET S2 = 0
0540 LET S8 = 0
0550 LET S9 = 0
0560 LET V9 = 0
0570 GOSUB 3010
0580 FOR J = 1 TO P
0590 FOR K = 1 TO Q
0600 FOR L = 1 TO R
0610 READ S
0620 IF S > 2E22 THEN 650
0630 LET N1 = N1 + 1
0640 GOTO 610
0650 GOSUB 3010
0660 IF G = 0 THEN 680
0670 GOSUB 3180
0680 READ S
0690 IF S > 2E22 THEN 720
0700 LET M8 = M8 + S/N1
```

Program 3. (cont.)

```
0710 GOTO 680
0720 GOSUB 3010
0730 IF G = 0 THEN 750
0740 GOSUB 3180
0750 READ S
0760 IF S > 2E22 THEN 800
0770 LET S9 = S9 + (S - M8)*(S - M8)
0780 LET V9 = V9 + (S - M8)*(S - M8)/(N1 - 1)
0790 GOTO 750
0800 PRINT "MEAN CELL A";J;"B";K;"C";L;" = ";M8;
0810 PRINT "VAR. CELL A";J;"B";K;"C";L;" = ";V9
0820 LET V9 = 0
0830 GOSUB 3110
0840 LET H = H + N1
0850 LET N1 = 0
0860 LET M5 = M5 + M8/R
0870 LET S8 = S8 + (M8 - M)*(M8 - M)
0880 LET M8 = 0
0890 LET G = G + 1
0900 GOSUB 3010
0910 GOSUB 3180
0920 NEXT L
0930 PRINT "MEAN A";J;"B";K;" = ";M5
0940 GOSUB 3110
0950 LET M2 = M2 + M5/Q
0960 LET S5 = S5 + (M5 - M)*(M5 - M)*R
0970 LET M5 = 0
0980 NEXT K
0990 LET S2 = S2 + (M2 - M)*(M2 - M)*Q*R
1000 PRINT "MEAN A";J;" = ";M2
1010 GOSUB 3110
1020 LET M2 = 0
1030 NEXT J
1040 REM CALCULATE MEAN EACH CELL (M8), MEAN EACH LEVEL
1050 REM FACTOR B (M3), SSB (S3), MEAN EACH LEVEL BC
1060 REM INTERACTION (M7), AND INTERMEDIATE VALUE
1070 REM SSBC (S7)
1080 LET H = 0
1090 LET H1 = 0
1100 LET H2 = 0
1110 LET M3 = 0
1120 LET M7 = 0
1130 LET M8 = 0
1140 LET N1 = 0
1150 LET S3 = 0
1160 LET S7 = 0
1170 FOR K = 1 TO Q
1180 FOR L = 1 TO R
1190 FOR J = 1 TO P
1200 GOSUB 3010
1210 IF H = 0 THEN 1240
1220 LET G = H
1230 GOSUB 3180
1240 IF H1 = 0 THEN 1270
1250 LET G = H1
1260 GOSUB 3180
1270 IF H2 = 0 THEN 1300
1280 LET G = H2
1290 GOSUB 3180
1300 READ S
1310 IF S > 2E22 THEN 1340
1320 LET N1 = N1 + 1
1330 GOTO 1300
1340 GOSUB 3010
1350 IF H = 0 THEN 1380
1360 LET G = H
1370 GOSUB 3180
1380 IF H1 = 0 THEN 1410
1390 LET G = H1
1400 GOSUB 3180
1410 IF H2 = 0 THEN 1440
```

Program 3. (cont.)

```
1420 LET G = H2
1430 GOSUB 3180
1440 READ S
1450 IF S > 2E22 THEN 1480
1460 LET M8 = M8 + S/N1
1470 GOTO 1440
1480 LET H = H + Q*R
1490 LET M7 = M7 + M8/P
1500 LET N1 = 0
1510 LET M8 = 0
1520 NEXT J
1530 LET S7 = S7 + (M7 - M)*(M7 - M)*P
1540 LET M3 = M3 + M7/R
1550 PRINT "MEAN B";K;"C";L;" = ";M7
1560 GOSUB 3110
1570 LET M7 = 0
1580 LET H = 0
1590 LET H1 = H1 + 1
1600 NEXT L
1610 PRINT "MEAN B";K;" = ";M3
1620 GOSUB 3110
1630 LET S3 = S3 + (M3 - M)*(M3 - M)*P*R
1640 LET M3 = 0
1650 LET H = 0
1660 LET H1 = 0
1670 LET H2 = H2 + R
1680 NEXT K
1690 REM CALCULATE MEAN EACH CELL (M8), MEAN EACH LEVEL
1700 REM FACTOR C (M4), SSC (S4), MEAN EACH LEVEL AC
1710 REM INTERACTION (M6), SSAC (S6), FINAL VALUES FOR
1720 REM SSAB (S5), SSAC (S6), SSBC (S7), SSABC (S8),
1730 REM AND SSWITHIN CELL (S9)
1740 LET H = 0
1750 LET H1 = 0
1760 LET M4 = 0
1770 LET M6 = 0
1780 LET M8 = 0
1790 LET N1 = 0
1800 LET S4 = 0
1810 LET S6 = 0
1820 FOR L = 1 TO R
1830 FOR J = 1 TO P
1840 FOR K = 1 TO Q
1850 GOSUB 3010
1860 IF H = 0 THEN 1890
1870 LET G = H
1880 GOSUB 3180
1890 IF H1 = 0 THEN 1920
1900 LET G = H1
1910 GOSUB 3180
1920 READ S
1930 IF S > 2E22 THEN 1960
1940 LET N1 = N1 + 1
1950 GOTO 1920
1960 GOSUB 3010
1970 IF H = 0 THEN 2000
1980 LET G = H
1990 GOSUB 3180
2000 IF H1 = 0 THEN 2030
2010 LET G = H1
2020 GOSUB 3180
2030 READ S
2040 IF S > 2E22 THEN 2070
2050 LET M8 = M8 + S/N1
2060 GOTO 2030
2070 LET H = H + R
2080 LET M6 = M6 + M8/Q
2090 LET N1 = 0
2100 LET M8 = 0
2110 NEXT K
2120 LET S6 = S6 + (M6 - M)*(M6 - M)*Q - S2/P/R
```

Program 3. (cont.)

```
2130 LET M4 = M4 + M6/P
2140 PRINT "MEAN A";J;"C";L;" = ";M6
2150 GOSUB 3110
2160 LET M6 = 0
2170 NEXT J
2180 PRINT "MEAN C";L;" = ";M4
2190 GOSUB 3110
2200 LET S4 = S4 + (M4 - M)*(M4 - M)*P*Q
2210 LET M4 = 0
2220 LET H = 0
2230 LET H1 = H1 + 1
2240 NEXT L
2250 LET S5 = S5 - S2 - S3
2260 LET S6 = S6 - S4
2270 LET S7 = S7 - S3 - S4
2280 LET S8 = S8 - S2 - S3 - S4 - S5 - S6 - S7
2290 LET S9 = S9/(N - P*Q*R)/P/Q/R*N2
2300 REM CREATE F-TABLE
2310 LET T = 13
2320 GOSUB 3110
2330 PRINT
2340 PRINT TAB(9);"SS";TAB(22);"DF";TAB(35);"MS";TAB(48);"F"
2350 PRINT "A";TAB(9);S2;TAB(22);P - 1;TAB(35);S2/(P - 1);
2360 PRINT TAB(48);S2/(P - 1)/S9
2370 PRINT "B";TAB(9);S3;TAB(22);Q - 1;TAB(35);S3/(Q - 1);
2380 PRINT TAB(48);S3/(Q - 1)/S9
2390 PRINT "C";TAB(9);S4;TAB(22);R - 1;TAB(35);S4/(R - 1);
2400 PRINT TAB(48);S4/(R - 1)/S9
2410 PRINT "AB";TAB(9);S5;TAB(22);(P - 1)*(Q - 1);TAB(35);
2420 PRINT S5/(P - 1)/(Q - 1);TAB(48);S5/(P - 1)/(Q - 1)/S9
2430 PRINT "AC";TAB(9);S6;TAB(22);(P - 1)*(R - 1);TAB(35);
2440 PRINT S6/(P - 1)/(R - 1);TAB(48);S6/(P - 1)/(R - 1)/S9
2450 PRINT "BC";TAB(9);S7;TAB(22);(Q - 1)*(R - 1);TAB(35);
2460 PRINT S7/(Q - 1)/(R - 1);TAB(48);S7/(Q - 1)/(R - 1)/S9
2470 PRINT "ABC";TAB(9);S8;TAB(22);(P - 1)*(Q - 1)*(R - 1);TAB(35);
2480 PRINT S8/(P - 1)/(Q - 1)/(R - 1);TAB(48);S8/(P-1)/(Q-1)/(R-1)/S9
2490 PRINT "W.CEL";TAB(22);N - P*Q*R;TAB(35);S9
2500 GOTO 9999
3000 REM RESTORE DATA POINTER AND ADVANCE DATA POINTER
3010 RESTORE
3020 READ S,S,S
3030 RETURN
3040 REM ADVANCE DATA POINTER
3050 FOR L2 = 1 TO L1
3060 READ S
3070 IF S > 3E24 THEN 3090
3080 NEXT L2
3090 RETURN
3100 REM SUBROUTINE TO STOP EXECUTION UNTIL DATA ARE TRANSCRIBED
3110 LET T = T + 1
3120 IF T <= 12 THEN RETURN
3130 PRINT "ENTER 1 FOLLOWED BY CARRIAGE RETURN TO CONTINUE."
3140 INPUT R1
3150 LET T = 0
3160 RETURN
3170 REM ADVANCE POINTER ONE OR MORE SUBGROUPS
3180 FOR L3 = 1 TO G
3190 READ S
3200 IF S > 2E22 THEN 3220
3210 GOTO 3190
3220 NEXT L3
3230 RETURN
9998 DATA 3E24
9999 END
```

Program 4. Completely randomized analysis of variance, CR-K; and randomized block analysis of variance, RB-K (requires equal n in all cells).

```
0010 REM THIS PROGRAM IS FOR AN INDEPENDENT GROUPS ONE-WAY
0020 REM ANALYSIS OF VARIANCE (COMPLETELY RANDOMIZED ANALYSIS
0030 REM OF VARIANCE, CR-K) AND A TWO-WAY ANALYSIS OF VARIANCE
0040 REM WITHOUT REPLICATIONS (RANDOMIZED BLOCK, RB-K)
0050 REM SEE APPROPRIATE EXAMPLE FOR ENTERING DATA.
0060 PRINT
0070 READ A, N, T
0080 REM CALCULATE GRAND MEAN
0090 LET M = 0
0100 FOR I = 1 TO N*T
0110 READ Q
0120 LET M = M + Q/(N*T)
0130 NEXT I
0140 REM CALCULATE MEAN FOR TREAT. (M1), SSTOT. (S1), SSBET.
0150 REM TREAT. (S2), VAR. FOR TREAT. (S3), AND SSWITH. CELL (S4).
0160 GOSUB 3010
0170 LET M1 = 0
0180 LET S1 = 0
0190 LET S2 = 0
0200 LET S3 = 0
0210 LET S4 = 0
0220 LET T1 = 0
0230 LET L = 0
0240 FOR J = 1 TO T
0250 FOR I = 1 TO N
0260 READ Q
0270 LET M1 = M1 + Q/N
0280 NEXT I
0290 PRINT "MEAN TREAT. ";J;" = ";M1;
0300 GOSUB 3120
0310 GOSUB 3010
0320 IF L = 0 THEN 350
0330 LET H = L
0340 GOSUB 3050
0350 FOR I = 1 TO N
0360 READ Q
0370 LET S1 = S1 + (Q - M)*(Q - M)
0380 LET S3 = S3 + (Q - M1)*(Q - M1)/(N - 1)
0390 LET S4 = S4 + (Q - M1)*(Q - M1)
0400 NEXT I
0410 PRINT "VAR. TREAT. ";J;" = ";S3
0420 GOSUB 3110
0430 LET S3 = 0
0440 LET S2 = S2 + (M1 - M)*(M1 - M)*N
0450 LET M1 = 0
0460 LET L = L + N
0470 LET H = L
0480 GOSUB 3010
0490 GOSUB 3050
0500 NEXT J
0510 IF A = 0 THEN 850
0520 REM CALC. MEAN FOR BLOCK (M2), SSBLOCK (S6), VAR. FOR BLOCK (S5)
0530 LET L = 0
0540 LET S6 = 0
0550 LET S5 = 0
0560 LET M2 = 0
0570 GOSUB 3010
0580 IF L = 0 THEN 610
0590 LET H = L
0600 GOSUB 3050
0610 FOR J = 1 TO T
0620 READ Q
0630 LET M2 = M2 + Q/T
0640 LET H = N - 1
0650 GOSUB 3050
0660 NEXT J
0670 GOSUB 3010
0680 IF L = 0 THEN 710
0690 LET H = L
0700 GOSUB 3050
```

Program 4. (cont.)

```
0710 FOR J = 1 TO T
0720 READ Q
0730 LET S5 = S5 + (Q - M2)*(Q - M2)/(T - 1)
0740 LET H = N - 1
0750 GOSUB 3050
0760 NEXT J
0770 PRINT "MEAN BLOCK ";L + 1;" = ";M2;
0780 PRINT "VAR. BLOCK ";L + 1;" = ";S5
0790 GOSUB 3110
0800 LET S6 = S6 + (M2 - M)*(M2 - M)*T
0810 LET L = L + 1
0820 IF L <> N THEN 550
0830 LET S7 = S4 - S6
0840 REM CALCULATE DEGREES OF FREEDOM
0850 LET D1 = N*T - 1
0860 LET D2 = T - 1
0870 LET D4 = (N - 1)*T
0880 LET D6 = N - 1
0890 LET D7 = (N - 1)*(T - 1)
0900 REM PRINT F TABLE
0910 LET T1 = 13
0920 GOSUB 3120
0930 PRINT TAB(9);"SS";TAB(22);"DF";TAB(35);"MS";TAB(48);"F"
0940 PRINT "TOTAL";TAB(9);S1;TAB(22);D1
0950 IF A = 0 THEN GOSUB 1010
0960 GOSUB 1110
0970 GOSUB 1050
0980 GOSUB 1140
0990 PRINT " REMAIN";TAB(9);S7;TAB(22);D7;TAB(35);S7/D7
1000 GOTO 1160
1010 GOSUB 1050
1020 GOSUB 1070
1030 GOSUB 1090
1040 GOTO 1160
1050 PRINT " TREAT.";TAB(9);S2;TAB(22);D2;TAB(35);S2/D2;TAB(48);
1060 RETURN
1070 PRINT (S2/D2)/(S4/D4)
1080 RETURN
1090 PRINT " W. CEL";TAB(9);S4;TAB(22);D4;TAB(35);S4/D4
1100 RETURN
1110 PRINT " BLOCKS";TAB(9);S6;TAB(22);D6;TAB(35);S6/D6;
1120 PRINT TAB(48);(S6/D6)/(S7/D7)
1130 RETURN
1140 PRINT (S2/D2)/(S7/D7)
1150 RETURN
1160 PRINT "TOTAL";TAB(9);S2 + S4;TAB(22);D1
1170 GOTO 9999
3000 REM RESTORE DATA POINTER AND ADVANCE TO FIRST DATA POINT
3010 RESTORE
3020 READ Q, Q, Q
3030 RETURN
3040 REM SUBROUTINE TO MOVE DATA POINTER TO NEXT OBSERVATION
3050 FOR L1 = 1 TO H
3060 READ Q
3070 IF Q > 2E22 THEN 3090
3080 NEXT L1
3090 RETURN
3100 REM SUBROUTINE TO STOP EXECUTION UNTIL DATA ARE TRANSCRIBED
3110 LET T1 = T1 + 1
3120 IF T1 < 10 THEN RETURN
3130 PRINT
3140 PRINT "ENTER 1 FOLLOWED BY CARRIAGE RETURN TO CONTINUE."
3150 INPUT Q
3160 LET T1 = 0
3170 RETURN
9998 DATA 9E99
9999 END
```

Program 5. Completely randomized factorial analysis of variance, CRF-JK; split-plot analysis of variance, SPF-J.K; and randomized block factorial analysis of variance, RBF-.JK (requires equal n in all cells).

```
0010 REM THIS PROGRAM IS FOR AN INDEPENDENT GROUPS
0020 REM TWO-WAY ANALYSIS OF VARIANCE (COMPLETELY RANDOMIZED
0030 REM FACTORIAL ANALYSIS OF VARIANCE (CRF-JK.)), SPLIT-
0040 REM PLOT ANALYSIS OF VARIANCE (SPF-J.K), AND RANDOMIZED
0050 REM BLOCK FACTORIAL ANALYSIS OF VARIANCE (RBF-.JK).
0060 PRINT
0070 REM SEE APPROPRIATE EXAMPLE FOR ENTERING DATA.
0080 READ A, N, P, Q
0090 REM CALCULATE GRAND MEAN
0100 GOSUB 3010
0110 LET M = 0
0120 FOR I = 1 TO N*P*Q
0130 READ R
0140 LET M = M + R/(N*P*Q)
0150 NEXT I
0160 REM CALCULATE SSTOTAL (S1), MEANS FOR EACH CELL (M2), VAR. FOR
0170 REM EACH CELL (V5), SSWITHIN CELLS (S5), MEANS FOR LEVELS OF
0180 REM FACTOR A (M1), SSA (S2), AND INTERMEDIATE VALUE FOR SSAB (S4).
0190 GOSUB 3010
0200 LET M1 = 0
0210 LET M2 = 0
0220 LET S1 = 0
0230 LET S2 = 0
0240 LET S4 = 0
0250 LET S5 = 0
0260 LET V2 = 0
0270 LET V5 = 0
0280 LET T = 0
0290 LET L = 0
0300 FOR J = 1 TO P
0310 FOR K = 1 TO Q
0320 FOR I = 1 TO N
0330 READ R
0340 LET S1 = S1 + (R - M)*(R - M)
0350 LET M1 = M1 + R/(N*Q)
0360 LET M2 = M2 + R/N
0370 NEXT I
0380 GOSUB 3010
0390 IF L<> 0 THEN GOSUB 3050
0400 FOR I = 1 TO N
0410 READ R
0420 LET S5 = S5 + (R - M2)*(R - M2)
0430 LET V5 = V5 + (R - M2)*(R - M2)/(N - 1)
0440 NEXT I
0450 PRINT "MEAN A ";J;" B ";K;" = ";M2;"VAR. A ";J;" B ";K;" = ";V5
0460 GOSUB 3100
0470 LET S4 = S4 + (M2 - M)*(M2 - M)*N
0480 LET M2 = 0
0490 LET V5 = 0
0500 LET L = L + N
0510 GOSUB 3010
0520 GOSUB 3050
0530 NEXT K
0540 LET S2 = S2 + (M1 - M)*(M1 - M)*N*Q
0550 PRINT "MEAN LEVEL A ";J;" = ";M1
0560 GOSUB 3100
0570 LET M1 = 0
0580 LET V2 = 0
0590 NEXT J
0600 REM CALCULATE MEANS FOR LEVELS OF FACTOR B (M3),
0610 REM SSB (S3), AND COMPLETED VALUE FOR SSAB (S4)
0620 LET L = 0
0630 LET M3 = 0
0640 LET S3 = 0
0650 FOR K = 1 TO Q
0660 GOSUB 3010
0670 IF L = 0 THEN 690
```

Program 5. (cont.)

```
0680 GOSUB 3050
0690 FOR J = 1 TO P
0700 FOR I = 1 TO N
0710 READ R
0720 LET M3 = M3 + R/(N*P)
0730 NEXT I
0740 GOSUB 3180
0750 NEXT J
0760 PRINT "MEAN B ";K;" = ";M3
0770 GOSUB 3110
0780 LET S3 = S3 + (M3 - M)*(M3 - M)*N*P
0790 LET L = L + N
0800 LET M3 = 0
0810 NEXT K
0820 LET S4 = S4 - S2 - S3
0830 IF A = 0 GOSUB 1830
0840 REM CALCULATE MEAN FOR EACH LEVEL OF A FOR A GIVEN
0850 REM BLOCK (M6), SSS.W.GRP. (S6), AND SSBXS.W.GRP. (S7).
0860 LET M6 = 0
0870 LET S6 = 0
0880 LET G = 0
0890 LET H = 0
0900 FOR J = 1 TO P
0910 FOR I = 1 TO N
0920 GOSUB 3010
0930 IF H = 0 THEN 960
0940 LET L = H
0950 GOSUB 3050
0960 IF G = 0 THEN 990
0970 LET L = G
0980 GOSUB 3050
0990 FOR K = 1 TO Q
1000 READ R
1010 LET M6 = M6 + R/Q
1020 LET L = N - 1
1030 GOSUB 3050
1040 NEXT K
1050 LET S6 = S6 + (M6 - M)*(M6 - M)*Q - S2/(N*P)
1060 PRINT "MEAN LEVEL A";J;"N";I;" = ";M6
1070 GOSUB 3110
1080 LET M6 = 0
1090 LET G = G + 1
1100 NEXT I
1110 LET G = 0
1120 LET H = H + N*Q
1130 NEXT J
1140 LET S7 = S5 - S6
1150 IF A = 1 GOSUB 2120
1160 REM CALCULATE MEAN BLOCK (M8), SSBLOCKS (S8),
1170 REM AND SSAXBLOCKS (S9).
1180 LET G = 0
1190 LET S8 = 0
1200 FOR I = 1 TO N
1210 LET M8 = 0
1220 GOSUB 3010
1230 IF G = 0 THEN 1260
1240 LET L = G
1250 GOSUB 3050
1260 FOR J = 1 TO P
1270 FOR K = 1 TO Q
1280 READ R
1290 LET M8 = M8 + R/(P*Q)
1300 LET L = N - 1
1310 GOSUB 3050
1320 NEXT K
1330 NEXT J
1340 LET S8 = S8 + (M8 - M)*(M8 - M)*P*Q
1350 PRINT "MEAN BLOCK ";I;" = ";M8
1360 GOSUB 3110
1370 LET G = G + 1
```

Program 5. (cont.)

```
1380 NEXT I
1390 LET S9 = S6 - S8
1400 REM CALCULATE MEAN FOR EACH LEVEL OF B FOR A GIVEN
1410 REM BLOCK (M0), SSBXBLOCKS (S0), AND SSABXBLOCKS (T1)
1420 LET H = 0
1430 LET S0 = 0
1440 LET M0 = 0
1450 FOR K = 1 TO Q
1460 FOR I = 1 TO N
1470 GOSUB 3010
1480 IF H = 0 THEN 1510
1490 LET L = H
1500 GOSUB 3050
1510 FOR J = 1 TO P
1520 READ R
1530 LET M0 = M0 + R/P
1540 LET L = N*Q - 1
1550 GOSUB 3050
1560 NEXT J
1570 LET S0 = S0 + (M0 - M)*(M0 - M)*P - S3/(Q*N) - S8/(Q*N)
1580 PRINT "MEAN LEVEL B";K;"N";I;" = ";M0
1590 GOSUB 3110
1600 LET H = H + 1
1610 LET M0 = 0
1620 NEXT I
1630 NEXT K
1640 LET T1 = S5 - S8 - S9 - S0
1650 GOSUB 1930
1660 LET D8 = N - 1
1670 LET D9 = D8*(P - 1)
1680 LET D0 = D8*(Q - 1)
1690 LET E1 = D9*(Q - 1)
1700 GOSUB 2000
1710 PRINT "  BLOCK";TAB(9);S8;TAB(22);D8;TAB(35);S8/D8;TAB(48);
1720 PRINT (S8/D8)/((S9 + S0 + T1)/(D9 + D0 + E1))
1730 GOSUB 2040
1740 PRINT (S2/D2)/(S9/D9)
1750 PRINT "  AXBLK";TAB(9);S9;TAB(22);D9;TAB(35);S9/D9
1760 GOSUB 2060
1770 PRINT (S3/D3)/(S0/D0)
1780 PRINT "  BXBLK";TAB(9);S0;TAB(22);D0;TAB(35);S0/D0
1790 GOSUB 2080
1800 PRINT (S4/D4)/(T1/E1)
1810 PRINT "  ABXBK";TAB(9);T1;TAB(22);E1;TAB(35);T1/E1
1820 GOTO 2100
1830 GOSUB 1930
1840 GOSUB 2000
1850 GOSUB 2040
1860 PRINT (S2/D2)/(S5/D5)
1870 GOSUB 2060
1880 PRINT (S3/D3)/(S5/D5)
1890 GOSUB 2080
1900 PRINT (S4/D4)/(S5/D5)
1910 PRINT "  W.CEL";TAB(9);S5;TAB(22);D5;TAB(35);S5/D5
1920 GOTO 2100
1930 GOSUB 2300
1940 LET D1 = P*Q*N - 1
1950 LET D2 = P - 1
1960 LET D3 = Q - 1
1970 LET D4 = D2*D3
1980 LET D5 = P*Q*(N - 1)
1990 RETURN
2000 PRINT
2010 PRINT TAB(9);"SS";TAB(22);"DF";TAB(35);"MS";TAB(48);"F"
2020 PRINT "TOTAL";TAB(9);S1;TAB(22);D1
2030 RETURN
2040 PRINT "  A";TAB(9);S2;TAB(22);D2;TAB(35);S2/D2;TAB(48);
2050 RETURN
2060 PRINT "  B";TAB(9);S3;TAB(22);D3;TAB(35);S3/D3;TAB(48);
2070 RETURN
```

Program 5. (cont.)

```
2080 PRINT "   AB";TAB(9);S4;TAB(22);D4;TAB(35);S4/D4;TAB(48);
2090 RETURN
2100 PRINT "TOTAL";TAB(9);S2 + S3 + S4 + S5;TAB(22);D1
2110 GOTO 9999
2120 GOSUB 1930
2130 GOSUB 2270
2140 GOSUB 2270
2150 GOSUB 2000
2160 PRINT " BET.";TAB(9);S2 + S6;TAB(22);D2 + D6
2170 GOSUB 2040
2180 PRINT (S2/D2)/(S6/D6)
2190 PRINT "  SWG";TAB(9);S6;TAB(22);D6;TAB(35);S6/D6
2200 PRINT " WITH";TAB(9);S3 + S4 + S7;TAB(22);D3 + D4 + D7
2210 GOSUB 2060
2220 PRINT (S3/D3)/(S7/D7)
2230 GOSUB 2080
2240 PRINT (S4/D4)/(S7/D7)
2250 PRINT "  BXSWG";TAB(9);S7;TAB(22);D7;TAB(35);S7/D7
2260 GOTO 2100
2270 LET D6 = P*(N - 1)
2280 LET D7 = D6*(Q - 1)
2290 RETURN
2300 LET T = 13
2310 GOSUB 3110
2320 LET T2 = S2 + S3 + S4 + S5
2330 RETURN
2999 END
3000 REM RESTORE DATA POINTER AND ADVANCE DATA POINTER
3010 RESTORE
3020 READ R, R, R, R
3030 RETURN
3040 REM ADVANCE DATA POINTER
3050 FOR L1 = 1 TO L
3060 READ R
3070 IF R > 2E22 THEN 3090
3080 NEXT L1
3090 RETURN
3100 REM SUBROUTINE TO STOP EXECUTION UNTIL DATA ARE TRANSCRIBED
3110 LET T = T + 1
3120 IF T <= 12 THEN RETURN
3130 PRINT "ENTER 1 FOLLOWED BY CARRIAGE RETURN TO CONTINUE."
3140 INPUT R1
3150 LET T = 0
3160 RETURN
3170 REM SUBROUTINE TO ADVANCE DATA POINTER TO NEXT LEVEL OF FACTOR B
3180 FOR L1 = 1 TO Q - 1
3190 FOR I = 1 TO N
3200 READ R
3210 IF R > 2E22 THEN 3240
3220 NEXT I
3230 NEXT L1
3240 RETURN
9998 DATA 2E23
9999 END
```

Program 6. Completely randomized factorial analysis of variance, CRF-JKL; split-plot analysis of variance, SPF-JK.L; split-plot analysis of variance, SPF-J.KL; and randomized block factorial analysis of variance, RBF-.JKL (requires equal n in all cells).

```
0010 REM THIS PROGRAM IS FOR AN INDEPENDENT GROUPS THREE-
0020 REM WAY ANALYSIS OF VARIANCE (COMPLETELY RANDOMIZED
0030 REM FACTORIAL ANALYSIS OF VARIANCE (CRF-JKL)), SPLIT-
0040 REM PLOT ANALYSIS OF VARIANCE (ONE REPEATED MEASURE,
0050 REM SPF-JK.L), SPLIT PLOT ANALYSIS OF VARIANCE (TWO
0060 REM REPEATED MEASURES, SPF-J.KL), AND RANDOMIZED
0070 REM BLOCK FACTORIAL ANALYSIS OF VARIANCE (RBF-.JKL)
0080 PRINT
0090 PRINT
0100 REM SEE APPROPRIATE EXAMPLE FOR ENTERING DATA
0110 READ A, N, P, Q, R
0120 REM CALCULATE GRAND MEAN
0130 GOSUB 3010
0140 LET M = 0
0150 FOR I = 1 TO N*P*Q*R
0160 READ S
0170 LET M = M + S/(N*P*Q*R)
0180 NEXT I
0190 REM CALCULATE MEAN EACH LEVEL FACTOR B (M3), AND SSB (S3)
0200 LET S3 = 0
0210 LET M3 = 0
0220 LET G = 0
0230 GOSUB 3010
0240 FOR K = 1 TO Q
0250 FOR J = 1 TO P
0260 FOR L = 1 TO R
0270 FOR I = 1 TO N
0280 READ S
0290 LET M3 = M3 + S/(N*P*R)
0300 NEXT I
0310 NEXT L
0320 LET H = N*R*(Q - 1)
0330 GOSUB 3050
0340 NEXT J
0350 LET S3 = S3 + (M3 - M)*(M3 - M)*N*P*R
0360 PRINT "MEAN B";K;" = ";M3
0370 GOSUB 3110
0380 LET M3 = 0
0390 LET G = G + N*R
0400 LET H = G
0410 GOSUB 3010
0420 GOSUB 3050
0430 NEXT K
0440 REM CALCULATE SSTOTAL (S1), MEAN EACH LEVEL
0450 REM FACTOR A (M2), SSA (S2), MEAN EACH LEVEL
0460 REM AB INTERACTION (M5), SSAB (S5), MEAN EACH LEVEL
0470 REM ABC INTERACTION (M8), VAR. EACH LEVEL ABC (V9),
0480 REM AND SSWITHIN CELLS (S9)
0490 LET S1 = 0
0500 LET S2 = 0
0510 LET S5 = 0
0520 LET S9 = 0
0530 LET M2 = 0
0540 LET M5 = 0
0550 LET M8 = 0
0560 LET V9 = 0
0570 LET G1 = 0
0580 LET G2 = 0
0590 FOR J = 1 TO P
0600 FOR K = 1 TO Q
0610 FOR L = 1 TO R
0620 GOSUB 3010
0630 IF G2 = 0 THEN 660
0640 LET H = G2
0650 GOSUB 3050
0660 IF G1 <> 0 THEN 750
0670 FOR I = 1 TO N
```

Program 6. (cont.)

```
0680 READ S
0690 LET M2 = M2 + S/(N*Q*R)
0700 LET M5 = M5 + S/(N*R)
0710 LET M8 = M8 + S/N
0720 NEXT I
0730 LET G1 = 1
0740 GOTO 620
0750 FOR I = 1 TO N
0760 READ S
0770 LET S1 = S1 + (S - M)*(S - M)
0780 LET S9 = S9 + (S - M8)*(S - M8)
0790 LET V9 = V9 + (S - M8)*(S - M8)/(N - 1)
0800 NEXT I
0810 PRINT "MEAN CELL A";J;"B";K;"C";L;" = ";M8;
0820 PRINT "VAR. CELL A";J;"B";K;"C";L;" = ";V9
0830 LET V9 = 0
0840 GOSUB 3110
0850 LET M8 = 0
0860 LET G1 = 0
0870 LET G2 = G2 + N
0880 NEXT L
0890 LET S5 = S5 + (M5 - M)*(M5 - 5)*N*R - S3/(P*Q)
0900 PRINT "MEAN A";J;"B";K;" = ";M5
0910 GOSUB 3110
0920 LET M5 = 0
0930 NEXT K
0940 LET S2 = S2 + (M2 - M)*(M2 - M)*N*Q*R
0950 PRINT "MEAN A";J;" = ";M2
0960 GOSUB 3110
0970 LET M2 = 0
0980 NEXT J
0990 LET S5 = S5 - S2
1000 REM CALCULATE MEAN EACH LEVEL FACTOR C (M4), AND SSC (S4)
1010 LET G = 0
1020 LET S4 = 0
1030 LET M4 = 0
1040 GOSUB 3010
1050 FOR L = 1 TO R
1060 FOR J = 1 TO P
1070 FOR K = 1 TO Q
1080 FOR I = 1 TO N
1090 READ S
1100 LET M4 = M4 + S/(N*P*Q)
1110 NEXT I
1120 LET H = N*(R - 1)
1130 GOSUB 3050
1140 NEXT K
1150 NEXT J
1160 LET S4 = S4 + (M4 - M)*(M4 - M)*N*P*Q
1170 PRINT "MEAN C";L;" = ";M4
1180 GOSUB 3110
1190 LET M4 = 0
1200 GOSUB 3010
1210 LET G = G + N
1220 LET H = G
1230 GOSUB 3050
1240 NEXT L
1250 REM CALCULATE MEAN EACH LEVEL AC (M6), AND SSAC (S6)
1260 LET M6 = 0
1270 LET S6 = 0
1280 LET G = 0
1290 LET G1 = 0
1300 GOSUB 3010
1310 FOR J = 1 TO P
1320 FOR L = 1 TO R
1330 FOR K = 1 TO Q
1340 FOR I = 1 TO N
1350 READ S
1360 LET M6 = M6 + S/(N*Q)
1370 NEXT I
1380 LET H = N*R - N
```

Program 6. (cont.)

```
1390 GOSUB 3050
1400 NEXT K
1410 LET S6 = S6 + (M6 - M)*(M6 - M)*N*Q - S2/(P*R) - S4/(P*R)
1420 PRINT "MEAN A";J;"C";L;" = ";M6
1430 GOSUB 3110
1440 LET M6 = 0
1450 LET G = G + N
1460 GOSUB 3010
1470 LET H = G
1480 GOSUB 3050
1490 IF G1 = 0 THEN 1520
1500 LET H = G1
1510 GOSUB 3050
1520 NEXT L
1530 LET G = 0
1540 LET G1 = G1 + N*Q*R
1550 GOSUB 3010
1560 LET H = G1
1570 GOSUB 3050
1580 NEXT J
1590 REM CALCULATE MEAN EACH LEVEL BC (M7), SSBC (S7) AND SSABC (S8)
1600 LET S7 = 0
1610 LET S8 = 0
1620 LET M7 = 0
1630 LET M8 = 0
1640 LET G = 0
1650 GOSUB 3010
1660 FOR K = 1 TO Q
1670 FOR L = 1 TO R
1680 FOR J = 1 TO P
1690 FOR I = 1 TO N
1700 READ S
1710 LET M7 = M7 + S/(N*P)
1720 LET M8 = M8 + S/N
1730 NEXT I
1740 LET G1 = P*Q*R
1750 LET S8 = S8 + (M8 - M)*(M8 - M)*N
1760 LET S8 = S8 - S2/G1 - S3/G1 - S4/G1 - S5/G1 - S6/G1
1770 LET M8 = 0
1780 LET H = N*R*Q - N
1790 GOSUB 3050
1800 NEXT J
1810 PRINT "MEAN B";K;"C";L;" = ";M7
1820 GOSUB 3110
1830 LET S7 = S7 + (M7 - M)*(M7 - M)*N*P - S3/(Q*R) - S4/(Q*R)
1840 LET M7 = 0
1850 LET G = G + N
1860 GOSUB 3010
1870 LET H = G
1880 GOSUB 3050
1890 NEXT L
1900 NEXT K
2000 LET S8 = S8 - S7
2010 LET D1 = N*P*Q*R - 1
2020 LET D2 = P - 1
2030 LET D3 = Q - 1
2040 LET D4 = R - 1
2050 LET D5 = D2*D3
2060 LET D6 = D2*D4
2070 LET D7 = D3*D4
2080 LET D8 = D5*D4
2090 LET D9 = P*Q*R*(N - 1)
2100 IF A = 0 THEN 4470
2110 REM CALCULATE SSS.W.GP (T1), FOR SPF-JK.L, MEAN
2120 REM EACH BLOCK (M0), SSBLOCKS (U1), SSS.W.GP (T3)
2130 REM FOR SPF-J.KL, SSCXS.W.GP (T2) FOR SPF-JK.L,
2140 REM SSBXS.W.GP (T4) FOR SPF-J.KL, SSAXBLOCKS
2150 REM (U2), AND SSABXBLOCKS (U5)
2160 LET M0 = 0
2170 LET M1 = 0
```

Program 6. (cont.)

```
2180 LET M3 = 0
2190 LET T1 = 0
2200 LET T3 = 0
2210 LET U1 = 0
2220 LET G = 0
2230 GOSUB 3010
2240 FOR I = 1 TO N
2250 FOR J = 1 TO P
2260 FOR K = 1 TO Q
2270 FOR L = 1 TO R
2280 READ S
2290 LET M0 = M0 + S/(P*Q*R)
2300 LET M1 = M1 + S/R
2310 LET M3 = M3 + S/(Q*R)
2320 LET H = N - 1
2330 GOSUB 3050
2340 NEXT L
2350 LET G1 = N*P*Q
2360 LET T1 = T1 + (M1 - M)*(M1 - M)*R - S2/G1 - S3/G1 - S5/G1
2380 LET M1 = 0
2390 NEXT K
2400 LET T3 = T3 + (M3 - M)*(M3 - M)*Q*R - S2/(N*P)
2410 LET M3 = 0
2420 NEXT J
2430 LET U1 = U1 + (M0 - M)*(M0 - M)*P*Q*R
2440 IF A < 3 THEN 2470
2450 PRINT "MEAN N";I;" = ";M0
2460 GOSUB 3110
2470 LET M0 = 0
2480 GOSUB 3010
2490 LET G = G + 1
2500 LET H = G
2510 GOSUB 3050
2520 NEXT I
2530 LET T2 = S9 - T1
2540 LET T4 = T1 - T3
2550 LET U2 = T3 - U1
2560 LET U5 = T1 - U1
2570 LET E1 = P*Q*(N - 1)
2580 LET E2 = P*Q*(R - 1)*(N - 1)
2590 IF A = 1 THEN 4250
2600 REM CALCULATE SSCXBLOCKS (U4), SSCXS.W.GP (T5) FOR SPF-J.KL,
2610 REM SSBCXS.W.GP (T6) FOR SPF-J.KL
2620 LET G = 0
2630 LET G1 = 0
2640 LET T5 = 0
2650 LET U4 = 0
2660 LET M4 = 0
2670 LET M5 = 0
2680 GOSUB 3010
2690 FOR L = 1 TO R
2700 FOR I = 1 TO N
2710 FOR J = 1 TO P
2720 FOR K = 1 TO Q
2730 READ S
2740 LET M4 = M4 + S/(P*Q)
2750 LET M5 = M5 + S/Q
2760 LET H = N*R - 1
2770 GOSUB 3050
2780 NEXT K
2790 LET G2 = N*P*R
2800 LET T5 = T5 + (M5 - M)*(M5 - M)*Q - S2/G2 - S4/G2 - S6/G2 - T3/G2
2810 LET M5 = 0
2820 NEXT J
2830 LET U4 = U4 + (M4 - M)*(M4 - M)*P*Q - S4/(N*R) - U1/(N*R)
2840 LET M4 = 0
2850 GOSUB 3010
2860 LET G = G + 1
2870 LET H = G
2880 GOSUB 3050
```

Program 6. (cont.)

```
2890 NEXT I
2900 GOSUB 3010
2910 LET H = G
2920 GOSUB 3050
2930 NEXT L
2940 LET T6 = S9 - T3 - T4 - T5
2950 LET E3 = P*(N - 1)
2960 LET E4 = P*(Q - 1)*(N - 1)
2970 LET E5 = P*(R - 1)*(N - 1)
2980 LET E6 = E4*(R - 1)
2990 GOTO 3170
3000 REM RESTORE DATA POINTER AND ADVANCE DATA POINTER
3010 RESTORE
3020 READ S, S, S, S, S
3030 RETURN
3040 REM ADVANCE DATA POINTER
3050 FOR L1 = 1 TO H
3060 READ S
3070 IF S > 2E22 THEN 3090
3080 NEXT L1
3090 RETURN
3100 REM SUBROUTINE TO STOP EXECUTION UNTIL DATA ARE TRANSCRIBED
3110 LET T = T + 1
3120 IF T <= 10 THEN RETURN
3130 PRINT "ENTER 1 FOLLOWED BY CARRIAGE RETURN TO CONTINUE."
3140 INPUT R1
3150 LET T = 0
3160 RETURN
3170 IF A = 2 THEN 4010
3180 REM CALCULATE SSBXBLOCKS (U3), SSBCXBLOCKS (U7),
3190 REM SSABCXBLOCKS (U8), SSABXBLOCKS (U5), AND SSACXBLOCKS (U6)
3200 LET G = 0
3210 LET G1 = 0
3220 LET G2 = 0
3230 LET M3 = 0
3240 LET M7 = 0
3250 LET U3 = 0
3260 LET U7 = 0
3270 GOSUB 3010
3280 FOR K = 1 TO Q
3290 FOR I = 1 TO N
3300 FOR L = 1 TO R
3310 FOR J = 1 TO P
3320 READ S
3330 LET M3 = M3 + S/(P*R)
3340 LET M7 = M7 + S/P
3350 LET H = N*Q*R - 1
3360 GOSUB 3050
3370 NEXT J
3380 LET G3 = (M7 - M)*(M7 - M)*P - S3/(N*Q*R) - S4/(N*Q*R) - S7/(N*Q*R)
3390 LET U7 = U7 + G3 - U1/(N*Q*R) - U4/(N*Q*R)
3400 LET M7 = 0
3410 GOSUB 3010
3420 LET G = G + N
3430 LET H = G
3440 GOSUB 3050
3450 NEXT L
3460 LET U3 = U3 + (M3 - M)*(M3 - M)*P*R - S3/(N*Q) - U1/(N*Q)
3470 LET M3 = 0
3480 GOSUB 3010
3490 LET G1 = G1 + 1
3500 LET G = G1 + G2
3510 LET H = G
3520 GOSUB 3050
3530 NEXT I
3540 GOSUB 3010
3550 LET G1 = 0
3560 LET G2 = G2 + N*R
3570 LET G = G2
3580 LET H = G
3590 GOSUB 3050
```

Program 6. (cont.)

```
3600 NEXT K
3610 LET U7 = U7 - U3
3620 LET U6 = T5 + T3 - U2 - U4 - U1
3630 LET U5 = T1 - U2 - U3 - U1
3640 LET U8 = S9 - U2 - U3 - U4 - U5 - U6 - U7 - U1
3650 LET F1 = N - 1
3660 LET F2 = (P - 1)*(N - 1)
3670 LET F3 = (Q - 1)*(N - 1)
3680 LET F4 = (R - 1)*(N - 1)
3690 LET F5 = F2*(Q - 1)
3700 LET F6 = F2*(R - 1)
3710 LET F7 = F4*(Q - 1)
3720 LET F8 = F7*(P - 1)
3730 REM CREATE F-TABLE FOR APPROPRIATE ANOVAR
3740 REM F-TABLE FOR RBF-.JKL
3750 GOSUB 4800
3760 PRINT "  BLOCK";TAB(9);U1;TAB(22);F1;TAB(35);U1/F1;TAB(48);
3770 PRINT (U1/F1)/((U2+U3+U4+U5+U6+U7+U8)/(F2+F3+F4+F5+F6+F7+F8))
3780 GOSUB 4640
3790 PRINT (S2/D2)/(U2/F2)
3800 PRINT "  AXBLK";TAB(9);U2;TAB(22);F2;TAB(35);U2/F2
3810 GOSUB 4660
3820 PRINT (S3/D3)/(U3/F3)
3830 PRINT "  BXBLK";TAB(9);U3;TAB(22);F3;TAB(35);U3/F3
3840 GOSUB 4680
3850 PRINT (S4/D4)/(U4/F4)
3860 PRINT "  CXBLK";TAB(9);U4;TAB(22);F4;TAB(35);U4/F4
3870 GOSUB 4700
3880 PRINT (S5/D5)/(U5/F5)
3890 PRINT "  ABXBK";TAB(9);U5;TAB(22);F5;TAB(35);U5/F5
3900 GOSUB 4720
3910 PRINT (S6/D6)/(U6/F6)
3920 PRINT "  ACXBK";TAB(9);U6;TAB(22);F6;TAB(35);U6/F6
3930 GOSUB 4740
3940 PRINT (S7/D7)/(U7/F7)
3950 PRINT "  BCXBK";TAB(9);U7;TAB(22);F7;TAB(35);U7/F7
3960 GOSUB 4760
3970 PRINT (S8/D8)/(U8/F8)
3980 PRINT "  ABCXB";TAB(9);U8;TAB(22);F8;TAB(35);U8/F8
3990 GOTO 4780
4000 REM SPF-J.KL
4010 GOSUB 4800
4020 PRINT " BET.";TAB(9);S2 + T3;TAB(22);D2 + E3
4030 GOSUB 4640
4040 PRINT (S2/D2)/(T3/E3)
4050 PRINT "  SWG.";TAB(9);T3;TAB(22);E3;TAB(35);T3/E3
4060 PRINT " WITH";TAB(9);S3 + S5 + T4 + S4 + S6 + T5 + S7 + S8 + T6;
4070 PRINT TAB(22);D3 + D5 + E4 + D4 + D6 + E5 + D7 + D8 + E6
4080 GOSUB 4660
4090 PRINT (S3/D3)/(T4/E4)
4100 GOSUB 4700
4110 PRINT (S5/D5)/(T4/E4)
4120 PRINT "  BXSWG";TAB(9);T4;TAB(22);E4;TAB(35);T4/E4
4130 GOSUB 4680
4140 PRINT (S4/D4)/(T5/E5)
4150 GOSUB 4720
4160 PRINT (S6/D6)/(T5/E5)
4170 PRINT "  CXSWG";TAB(9);T5;TAB(22);E5;TAB(35);T5/E5
4180 GOSUB 4740
4190 PRINT (S7/D7)/(T6/E6)
4200 GOSUB 4760
4210 PRINT (S8/D8)/(T6/E6)
4220 PRINT "  BCXSG";TAB(9);T6;TAB(22);E6;TAB(35);T6/E6
4230 GOTO 4780
4240 REM SPF-JK.L
4250 GOSUB 4800
4260 PRINT " BET.";TAB(9);S2+S3+S5+T1;TAB(22);D2+D3+D5+E1
4270 GOSUB 4640
4280 PRINT (S2/D2)/(T1/E1)
4290 GOSUB 4660
4300 PRINT (S3/D3)/(T1/E1)
```

Program 6. (cont.)

```
4310 GOSUB 4700
4320 PRINT (S5/D5)/(T1/E1)
4330 PRINT "  SWG.";TAB(9);T1;TAB(22);E1;TAB(35);T1/E1
4340 PRINT " WITH";TAB(9);S4 + S6 + S7 + S8 + T2;TAB(22);
4350 PRINT D4 + D6 + D7 + D8 + E2
4360 GOSUB 4680
4370 PRINT (S4/D4)/(T2/E2)
4380 GOSUB 4720
4390 PRINT (S6/D6)/(T2/E2)
4400 GOSUB 4740
4410 PRINT (S7/D7)/(T2/E2)
4420 GOSUB 4760
4430 PRINT (S8/D8)/(T2/E2)
4440 PRINT "  CXSWG";TAB(9);T2;TAB(22);E2;TAB(35);T2/E2
4450 GOTO 4780
4460 REM F-TABLE CRF-JKL
4470 GOSUB 4800
4480 GOSUB 4640
4490 PRINT (S2/D2)/(S9/D9)
4500 GOSUB 4660
4510 PRINT (S3/D3)/(S9/D9)
4520 GOSUB 4680
4530 PRINT (S4/D4)/(S9/D9)
4540 GOSUB 4700
4550 PRINT (S5/D5)/(S9/D9)
4560 GOSUB 4720
4570 PRINT (S6/D6)/(S9/D9)
4580 GOSUB 4740
4590 PRINT (S7/D7)/(S9/D9)
4600 GOSUB 4760
4610 PRINT (S8/D8)/(S9/D9)
4620 PRINT "  W.CEL";TAB(9);S9;TAB(22);D9;TAB(35);S9/D9
4630 GOTO 4780
4640 PRINT "  A";TAB(9);S2;TAB(22);D2;TAB(35);S2/D2;TAB(48);
4650 RETURN
4660 PRINT "  B";TAB(9);S3;TAB(22);D3;TAB(35);S3/D3;TAB(48);
4670 RETURN
4680 PRINT "  C";TAB(9);S4;TAB(22);D4;TAB(35);S4/D4;TAB(48);
4690 RETURN
4700 PRINT "  AB";TAB(9);S5;TAB(22);D5;TAB(35);S5/D5;TAB(48);
4710 RETURN
4720 PRINT "  AC";TAB(9);S6;TAB(22);D6;TAB(35);S6/D6;TAB(48);
4730 RETURN
4740 PRINT "  BC";TAB(9);S7;TAB(22);D7;TAB(35);S7/D7;TAB(48);
4750 RETURN
4760 PRINT "  ABC";TAB(9);S8;TAB(22);D8;TAB(35);S8/D8;TAB(48);
4770 RETURN
4780 PRINT "TOTAL";TAB(9);S2+S3+S4+S5+S6+S7+S8+S9;TAB(22);D1
4790 GOTO 9999
4800 LET T = 13
4810 GOSUB 3120
4820 PRINT
4830 PRINT TAB(9);"SS";TAB(22);"DF";TAB(35);"MS";TAB(48);"F"
4840 PRINT "TOTAL";TAB(9);S1;TAB(22);D1
4850 RETURN
9998 DATA 2E23
9999 END
```

Program 7. Completely randomized analysis of variance (requires equal n in all cells), CR-K.

```
0010 REM THIS PROGRAM IS FOR AN INDEPENDENT GROUPS ONE-WAY
0020 REM ANALYSIS OF VARIANCE (COMPLETELY RANDOMIZED ANALYSIS OF
0030 REM VARIANCE, CR-K)
0050 REM SEE APPROPRIATE EXAMPLE FOR ENTERING DATA.
0060 PRINT
0070 READ A, N, T
0080 REM CALCULATE GRAND MEAN
0090 LET M = 0
0100 FOR I = 1 TO N*T
0110 READ Q
0120 LET M = M + Q/(N*T)
0130 NEXT I
0140 REM CALCULATE MEAN FOR TREAT.(M1), SSTOT.(S1), SSBET.
0150 REM TREAT. (S2), VAR. FOR TREAT.(S3), AND SSWITH. CELL(S4).
0160 GOSUB 3010
0170 LET M1 = 0
0180 LET S1 = 0
0190 LET S2 = 0
0200 LET S3 = 0
0210 LET S4 = 0
0220 LET T1 = 0
0230 LET L = 0
0240 FOR J = 1 TO T
0250 FOR I = 1 TO N
0260 READ Q
0270 LET M1 = M1 + Q/N
0280 NEXT I
0290 PRINT "MEAN TREAT. ";J;" = ";M1;
0300 GOSUB 3120
0310 GOSUB 3010
0320 IF L = 0 THEN 350
0330 LET H = L
0340 GOSUB 3050
0350 FOR I = 1 TO N
0360 READ Q
0370 LET S1 = S1 + (Q - M)*(Q - M)
0380 LET S3 = S3 + (Q - M1)*(Q - M1)/(N - 1)
0390 LET S4 = S4 + (Q - M1)*(Q - M1)
0400 NEXT I
0410 PRINT "VAR. TREAT. ";J;" = ";S3
0420 GOSUB 3110
0430 LET S3 = 0
0440 LET S2 = S2 + (M1 - M)*(M1 - M)*N
0450 LET M1 = 0
0460 LET L = L + N
0470 LET H = L
0480 GOSUB 3010
0490 GOSUB 3050
0500 NEXT J
0840 REM CALCULATE DEGREES OF FREEDOM
0850 LET D1 = N*T - 1
0860 LET D2 = T - 1
0870 LET D4 = (N - 1)*T
0880 LET D6 = N - 1
0890 LET D7 = (N - 1)*(T - 1)
0900 REM PRINT F TABLE
0910 LET T1 = 13
0920 GOSUB 3120
0930 PRINT TAB(9);"SS";TAB(22);"DF";TAB(35);"MS";TAB(48);"F"
0940 PRINT "TOTAL";TAB(9);S1;TAB(22);D1
1050 PRINT " TREAT.";TAB(9);S2;TAB(22);D2;TAB(35);S2/D2;TAB(48);
1070 PRINT (S2/D2)/(S4/D4)
1090 PRINT " W. CEL";TAB(9);S4;TAB(22);D4;TAB(35);S4/D4
1160 PRINT "TOTAL";TAB(9);S2 + S4;TAB(22);D1
1170 GOTO 9999
3000 REM RESTORE DATA POINTER AND ADVANCE TO FIRST DATA POINT
3010 RESTORE
3020 READ Q, Q, Q
3030 RETURN
3040 REM SUBROUTINE TO MOVE DATA POINTER TO NEXT OBSERVATION
3050 FOR L1 = 1 TO H
```

Program 7. (cont.)

```
3060 READ Q
3070 IF Q > 2E22 THEN 3090
3080 NEXT L1
3090 RETURN
3100 REM SUBROUTINE TO STOP EXECUTION UNTIL DATA ARE TRANSCRIBED
3110 LET T1 = T1 + 1
3120 IF T1 < 10 THEN RETURN
3130 PRINT
3140 PRINT "ENTER 1 FOLLOWED BY CARRIAGE RETURN TO CONTINUE."
3150 INPUT Q
3160 LET T1 = 0
3170 RETURN
9998 DATA 2E23
9999 END
```

Program 8. Randomized block analysis of variance (requires equal n in all cells), RB-K.

```
0010 REM THIS PROGRAM IS FOR A TWO-WAY ANALYSIS OF
0020 REM VARIANCE WITHOUT REPLICATIONS (RANDOMIZED
0030 REM BLOCK, RB-K)
0050 REM SEE APPROPRIATE EXAMPLE FOR ENTERING DATA.
0060 PRINT
0070 READ A, N, T
0080 REM CALCULATE GRAND MEAN
0090 LET M = 0
0100 FOR I = 1 TO N*T
0110 READ Q
0120 LET M = M + Q/(N*T)
0130 NEXT I
0140 REM CALCULATE MEAN FOR TREAT.(M1), SSTOT.(S1), SSBET.
0150 REM TREAT.(S2), VAR. FOR TREAT.(S3), AND SSWITH. CELL(S4).
0160 GOSUB 3010
0170 LET M1 = 0
0180 LET S1 = 0
0190 LET S2 = 0
0200 LET S3 = 0
0210 LET S4 = 0
0220 LET T1 = 0
0230 LET L = 0
0240 FOR J = 1 TO T
0250 FOR I = 1 TO N
0260 READ Q
0270 LET M1 = M1 + Q/N
0280 NEXT I
0290 PRINT "MEAN TREAT. ";J;" = ";M1;
0300 GOSUB 3120
0310 GOSUB 3010
0320 IF L = 0 THEN 350
0330 LET H = L
0340 GOSUB 3050
0350 FOR I = 1 TO N
0360 READ Q
0370 LET S1 = S1 + (Q - M)*(Q - M)
0380 LET S3 = S3 + (Q - M1)*(Q - M1)/(N - 1)
0390 LET S4 = S4 + (Q - M1)*(Q - M1)
0400 NEXT I
0410 PRINT "VAR. TREAT. ";J;" = ";S3
0420 GOSUB 3110
0430 LET S3 = 0
0440 LET S2 = S2 + (M1 - M)*(M1 - M)*N
0450 LET M1 = 0
0460 LET L = L + N
0470 LET H = L
0480 GOSUB 3010
0490 GOSUB 3050
0500 NEXT J
0510 REM CALCULATE MEAN FOR BLOCK (M2), SSBLOCK (S6), VAR.
0520 REM FOR BLOCK (S5).
0530 LET L = 0
0540 LET S6 = 0
0550 LET S5 = 0
0560 LET M2 = 0
0570 GOSUB 3010
0580 IF L = 0 THEN 610
0590 LET H = L
0600 GOSUB 3050
0610 FOR J = 1 TO T
0620 READ Q
0630 LET M2 = M2 + Q/T
0640 LET H = N - 1
0650 GOSUB 3050
0660 NEXT J
0670 GOSUB 3010
0680 IF L = 0 THEN 710
0690 LET H = L
0700 GOSUB 3050
0710 FOR J = 1 TO T
```

Program 8. (cont.)

```
0720 READ Q
0730 LET S5 = S5 + (Q - M2)*(Q - M2)/(T - 1)
0740 LET H = N - 1
0750 GOSUB 3050
0760 NEXT J
0770 PRINT "MEAN BLOCK ";L + 1;" = ";M2;
0780 PRINT "VAR. BLOCK ";L + 1;" = ";S5
0790 GOSUB 3110
0800 LET S6 = S6 + (M2 - M)*(M2 - M)*T
0810 LET L = L + 1
0820 IF L <> N THEN 550
0830 LET S7 = S4 - S6
0840 REM CALCULATE DEGREES OF FREEDOM
0850 LET D1 = N*T - 1
0860 LET D2 = T - 1
0870 LET D4 = (N - 1)*T
0880 LET D6 = N - 1
0890 LET D7 = (N - 1)*(T - 1)
0900 REM PRINT F TABLE
0910 LET T1 = 13
0920 GOSUB 3120
0930 PRINT TAB(9);"SS";TAB(22);"DF";TAB(35);"MS";TAB(48);"F"
0940 PRINT "TOTAL";TAB(9);S1;TAB(22);D1
1110 PRINT " BLOCKS";TAB(9);S6;TAB(22);D6;TAB(35);S6/D6;
1120 PRINT TAB(48);(S6/D6)/(S7/D7)
1130 PRINT " TREAT.";TAB(9);S2;TAB(22);D2;TAB(35);S2/D2;TAB(48);
1140 PRINT (S2/D2)/(S7/D7)
1150 PRINT " REMAIN";TAB(9);S7;TAB(22);D7;TAB(35);S7/D7
1160 PRINT "TOTAL";TAB(9);S2 + S4;TAB(22);D1
1170 GOTO 9999
3000 REM RESTORE DATA POINTER AND ADVANCE TO FIRST DATA POINT
3010 RESTORE
3020 READ Q, Q, Q
3030 RETURN
3040 REM SUBROUTINE TO MOVE DATA POINTER TO NEXT OBSERVATION
3050 FOR L1 = 1 TO H
3060 READ Q
3070 IF Q > 2E22 THEN 3090
3080 NEXT L1
3090 RETURN
3100 REM SUBROUTINE TO STOP EXECUTION UNTIL DATA ARE TRANSCRIBED
3110 LET T1 = T1 + 1
3120 IF T1 < 10 THEN RETURN
3130 PRINT
3140 PRINT "ENTER 1 FOLLOWED BY CARRIAGE RETURN TO CONTINUE."
3150 INPUT Q
3160 LET T1 = 0
3170 RETURN
9998 DATA 2E23
9999 END
```

Program 9. Completely randomized factorial analysis of variance (requires equal n in all cells), CRF-JK.

```
0010 REM THIS PROGRAM IS FOR AN INDEPENDENT GROUPS
0020 REM TWO-WAY ANALYSIS OF VARIANCE (COMPLETELY RANDOMIZED
0030 REM FACTORIAL ANALYSIS OF VARIANCE (CRF-JK.))
0060 PRINT
0070 REM SEE APPROPRIATE EXAMPLE FOR ENTERING DATA.
0080 READ A, N, P, Q
0090 REM CALCULATE GRAND MEAN
0100 GOSUB 3010
0110 LET M = 0
0120 FOR I = 1 TO N*P*Q
0130 READ R
0140 LET M = M + R/(N*P*Q)
0150 NEXT I
0160 REM CALCULATE SSTOTAL (S1), MEANS FOR EACH CELL (M2), VAR. FOR
0170 REM EACH CELL (V5), SSWITHIN CELLS (S5), MEANS FOR LEVELS OF
0180 REM FACTOR A(M1), SSA(S2), AND INTERMEDIATE VALUE FOR SSAB(S4).
0190 GOSUB 3010
0200 LET M1 = 0
0210 LET M2 = 0
0220 LET S1 = 0
0230 LET S2 = 0
0240 LET S4 = 0
0250 LET S5 = 0
0260 LET V2 = 0
0270 LET V5 = 0
0280 LET T = 0
0290 LET L = 0
0300 FOR J = 1 TO P
0310 FOR K = 1 TO Q
0320 FOR I = 1 TO N
0330 READ R
0340 LET S1 = S1 + (R - M)*(R - M)
0350 LET M1 = M1 + R/(N*Q)
0360 LET M2 = M2 + R/N
0370 NEXT I
0380 GOSUB 3010
0390 IF L<> 0 THEN GOSUB 3050
0400 FOR I = 1 TO N
0410 READ R
0420 LET S5 = S5 + (R - M2)*(R - M2)
0430 LET V5 = V5 + (R - M2)*(R - M2)/(N - 1)
0440 NEXT I
0450 PRINT "MEAN A ";J;" B ";K;" = ";M2;"VAR. A ";J;" B ";K;" = ";V5
0460 GOSUB 3100
0470 LET S4 = S4 + (M2 - M)*(M2 - M)*N
0480 LET M2 = 0
0490 LET V5 = 0
0500 LET L = L + N
0510 GOSUB 3010
0520 GOSUB 3050
0530 NEXT K
0540 LET S2 = S2 + (M1 - M)*(M1 - M)*N*Q
0550 PRINT "MEAN LEVEL A ";J;" = ";M1
0560 GOSUB 3100
0570 LET M1 = 0
0580 LET V2 = 0
0590 NEXT J
0600 REM CALCULATE MEANS FOR LEVELS OF FACTOR B (M3),
0610 REM SSB (S3), AND COMPLETED VALUE FOR SSAB (S4)'
0620 LET L = 0
0630 LET M3 = 0
0640 LET S3 = 0
0650 FOR K = 1 TO Q
0660 GOSUB 3010
0670 IF L = 0 THEN 690
0680 GOSUB 3050
0690 FOR J = 1 TO P
0700 FOR I = 1 TO N
0710 READ R
0720 LET M3 = M3 + R/(N*P)
0730 NEXT I
```

Program 9. (cont.)

```
0740 GOSUB 3180
0750 NEXT J
0760 PRINT "MEAN B ";K;" = ";M3
0770 GOSUB 3110
0780 LET S3 = S3 + (M3 - M)*(M3 - M)*N*P
0790 LET L = L + N
0800 LET M3 = 0
0810 NEXT K
0820 LET S4 = S4 - S2 - S3
0830 LET T = 13
0840 GOSUB 3110
0850 LET T2 = S2 + S3 + S4 + S5
0860 LET D1 = P*Q*N - 1
0870 LET D2 = P - 1
0880 LET D3 = Q - 1
0890 LET D4 = D2*D3
0900 LET D5 = P*Q*(N - 1)
0910 PRINT
0920 PRINT TAB(9);"SS";TAB(22);"DF";TAB(35);"MS";TAB(48);"F"
0930 PRINT "TOTAL";TAB(9);S1;TAB(22);D1
0940 PRINT "  A";TAB(9);S2;TAB(22);D2;TAB(35);S2/D2;TAB(48);
0950 PRINT (S2/D2)/(S5/D5)
0960 PRINT "  B";TAB(9);S3;TAB(22);D3;TAB(35);S3/D3;TAB(48);
0970 PRINT (S3/D3)/(S5/D5)
0980 PRINT "  AB";TAB(9);S4;TAB(22);D4;TAB(35);S4/D4;TAB(48);
0990 PRINT (S4/D4)/(S5/D5)
1000 PRINT "  W.CEL";TAB(9);S5;TAB(22);D5;TAB(35);S5/D5
1010 PRINT "TOTAL";TAB(9);S2 + S3 + S4 + S5;TAB(22);D1
1020 GOTO 9999
3000 REM RESTORE DATA POINTER AND ADVANCE DATA POINTER
3010 RESTORE
3020 READ R, R, R, R
3030 RETURN
3040 REM ADVANCE DATA POINTER
3050 FOR L1 = 1 TO L
3060 READ R
3070 IF R > 2E22 THEN 3090
3080 NEXT L1
3090 RETURN
3100 REM SUBROUTINE TO STOP EXECUTION UNTIL DATA ARE TRANSCRIBED
3110 LET T = T + 1
3120 IF T <= 12 THEN RETURN
3130 PRINT "ENTER 1 FOLLOWED BY CARRIAGE RETURN TO CONTINUE."
3140 INPUT R1
3150 LET T = 0
3160 RETURN
3170 REM SUBROUTINE TO ADVANCE DATA POINTER TO NEXT LEVEL OF FACTOR B
3180 FOR L1 = 1 TO Q - 1
3190 FOR I = 1 TO N
3200 READ R
3210 IF R > 2E22 THEN 3240
3220 NEXT I
3230 NEXT L1
3240 RETURN
9998 DATA 2E23
9999 END
```

Program 10. Split-plot analysis of variance (requires equal n in all cells), SPF-J.K.

```
0010 REM THIS PROGRAM IS FOR A SPLIT-PLOT ANALYSIS
0020 REM OF VARIANCE (SPF-J.K)
0060 PRINT
0070 REM SEE APPROPRIATE EXAMPLE FOR ENTERING DATA.
0080 READ A, N, P, Q
0090 REM CALCULATE GRAND MEAN
0100 GOSUB 3010
0110 LET M = 0
0120 FOR I = 1 TO N*P*Q
0130 READ R
0140 LET M = M + R/(N*P*Q)
0150 NEXT I
0160 REM CALCULATE SSTOTAL (S1), MEANS FOR EACH CELL (M2), VAR. FOR
0170 REM EACH CELL (V5), SSWITHIN CELLS (S5), MEANS FOR LEVELS OF
0180 REM FACTOR A(M1), SSA(S2), AND INTERMEDIATE VALUE FOR SSAB(S4).
0190 GOSUB 3010
0200 LET M1 = 0
0210 LET M2 = 0
0220 LET S1 = 0
0230 LET S2 = 0
0240 LET S4 = 0
0250 LET S5 = 0
0260 LET V2 = 0
0270 LET V5 = 0
0280 LET T = 0
0290 LET L = 0
0300 FOR J = 1 TO P
0310 FOR K = 1 TO Q
0320 FOR I = 1 TO N
0330 READ R
0340 LET S1 = S1 + (R - M)*(R - M)
0350 LET M1 = M1 + R/(N*Q)
0360 LET M2 = M2 + R/N
0370 NEXT I
0380 GOSUB 3010
0390 IF L<> 0 THEN GOSUB 3050
0400 FOR I = 1 TO N
0410 READ R
0420 LET S5 = S5 + (R - M2)*(R - M2)
0430 LET V5 = V5 + (R - M2)*(R - M2)/(N - 1)
0440 NEXT I
0450 PRINT "MEAN A ";J;" B ";K;" = ";M2;"VAR. A ";J;" B ";K;" = ";V5
0460 GOSUB 3100
0470 LET S4 = S4 + (M2 - M)*(M2 - M)*N
0480 LET M2 = 0
0490 LET V5 = 0
0500 LET L = L + N
0510 GOSUB 3010
0520 GOSUB 3050
0530 NEXT K
0540 LET S2 = S2 + (M1 - M)*(M1 - M)*N*Q
0550 PRINT "MEAN LEVEL A ";J;" = ";M1
0560 GOSUB 3100
0570 LET M1 = 0
0580 LET V2 = 0
0590 NEXT J
0600 REM CALCULATE MEANS FOR LEVELS OF FACTOR B (M3),
0610 REM SSB (S3), AND COMPLETED VALUE FOR SSAB (S4)'
0620 LET L = 0
0630 LET M3 = 0
0640 LET S3 = 0
0650 FOR K = 1 TO Q
0660 GOSUB 3010
0670 IF L = 0 THEN 690
0680 GOSUB 3050
0690 FOR J = 1 TO P
0700 FOR I = 1 TO N
0710 READ R
0720 LET M3 = M3 + R/(N*P)
0730 NEXT I
```

Program 10. (cont.)

```
0740 GOSUB 3180
0750 NEXT J
0760 PRINT "MEAN B ";K;" = ";M3
0770 GOSUB 3110
0780 LET S3 = S3 + (M3 - M)*(M3 - M)*N*P
0790 LET L = L + N
0800 LET M3 = 0
0810 NEXT K
0820 LET S4 = S4 - S2 - S3
0840 REM CALCULATE MEAN FOR EACH LEVEL OF A FOR A GIVEN
0850 REM BLOCK (M6), SSS.W.GRP. (S6), AND SSBXS.W.GRP. (S7).
0860 LET M6 = 0
0870 LET S6 = 0
0880 LET G = 0
0890 LET H = 0
0900 FOR J = 1 TO P
0910 FOR I = 1 TO N
0920 GOSUB 3010
0930 IF H = 0 THEN 960
0940 LET L = H
0950 GOSUB 3050
0960 IF G = 0 THEN 990
0970 LET L = G
0980 GOSUB 3050
0990 FOR K = 1 TO Q
1000 READ R
1010 LET M6 = M6 + R/Q
1020 LET L = N - 1
1030 GOSUB 3050
1040 NEXT K
1050 LET S6 = S6 + (M6 - M)*(M6 - M)*Q - S2/(N*P)
1060 PRINT "MEAN LEVEL A";J;"N";I;" = ";M6
1070 GOSUB 3110
1080 LET M6 = 0
1090 LET G = G + 1
1100 NEXT I
1110 LET G = 0
1120 LET H = H + N*Q
1130 NEXT J
1140 LET S7 = S5 - S6
1150 LET T = 13
1160 GOSUB 3110
1170 LET T2 = S2 + S3 + S4 + S5
1190 LET D1 = P*Q*N - 1
1200 LET D2 = P - 1
1210 LET D3 = Q - 1
1220 LET D4 = D2*D3
1230 LET D5 = P*Q*(N - 1)
1240 LET D6 = P*(N - 1)
1250 LET D7 = D6*(Q - 1)
1260 PRINT
1270 PRINT TAB(9);"SS";TAB(22);"DF";TAB(35);"MS";TAB(48);"F"
1280 PRINT "TOTAL";TAB(9);S1;TAB(22);D1
1290 PRINT " BET.";TAB(9);S2 + S6;TAB(22);D2 + D6
1300 PRINT "  A";TAB(9);S2;TAB(22);D2;TAB(35);S2/D2;TAB(48);
1310 PRINT (S2/D2)/(S6/D6)
1320 PRINT "  SWG";TAB(9);S6;TAB(22);D6;TAB(35);S6/D6
1330 PRINT " WITH";TAB(9);S3 + S4 + S7;TAB(22);D3 + D4 + D7
1340 PRINT "  B";TAB(9);S3;TAB(22);D3;TAB(35);S3/D3;TAB(48);
1350 PRINT (S3/D3)/(S7/D7)
1360 PRINT "  AB";TAB(9);S4;TAB(22);D4;TAB(35);S4/D4;TAB(48);
1370 PRINT (S4/D4)/(S7/D7)
1380 PRINT "  BXSWG";TAB(9);S7;TAB(22);D7;TAB(35);S7/D7
1390 PRINT "TOTAL";TAB(9);S2 + S3 + S4 + S5;TAB(22);D1
1400 GOTO 9999
1990 RETURN
2290 RETURN
2330 RETURN
3000 REM RESTORE DATA POINTER AND ADVANCE DATA POINTER
3010 RESTORE
3020 READ R, R, R, R
3030 RETURN
```

Program 10. (cont.)

```
3040 REM ADVANCE DATA POINTER
3050 FOR L1 = 1 TO L
3060 READ R
3070 IF R > 2E22 THEN 3090
3080 NEXT L1
3090 RETURN
3100 REM SUBROUTINE TO STOP EXECUTION UNTIL DATA ARE TRANSCRIBED
3110 LET T = T + 1
3120 IF T <= 12 THEN RETURN
3130 PRINT "ENTER 1 FOLLOWED BY CARRIAGE RETURN TO CONTINUE."
3140 INPUT R1
3150 LET T = 0
3160 RETURN
3170 REM SUBROUTINE TO ADVANCE DATA POINTER TO NEXT LEVEL OF FACTOR B
3180 FOR L1 = 1 TO Q - 1
3190 FOR I = 1 TO N
3200 READ R
3210 IF R > 2E22 THEN 3240
3220 NEXT I
3230 NEXT L1
3240 RETURN
9998 DATA 2E23
9999 END
```

Program 11. Randomized block factorial analysis of variance (requires equal n in all cells), RBF-.JK.

```
0010 REM THIS PROGRAM IS FOR A RANDOMIZED BLOCK
0020 REM FACTORIAL ANALYSIS OF VARIANCE (RBF-.JK)
0060 PRINT
0070 REM SEE APPROPRIATE EXAMPLE FOR ENTERING DATA.
0080 READ A, N, P, Q
0090 REM CALCULATE GRAND MEAN
0100 GOSUB 3010
0110 LET M = 0
0120 FOR I = 1 TO N*P*Q
0130 READ R
0140 LET M = M + R/(N*P*Q)
0150 NEXT I
0160 REM CALCULATE SSTOTAL (S1), MEANS FOR EACH CELL (M2), VAR. FOR
0170 REM EACH CELL (V5), SSWITHIN CELLS (S5), MEANS FOR LEVELS OF
0180 REM FACTOR A(M1), SSA(S2), AND INTERMEDIATE VALUE FOR SSAB(S4).
0190 GOSUB 3010
0200 LET M1 = 0
0210 LET M2 = 0
0220 LET S1 = 0
0230 LET S2 = 0
0240 LET S4 = 0
0250 LET S5 = 0
0260 LET V2 = 0
0270 LET V5 = 0
0280 LET T = 0
0290 LET L = 0
0300 FOR J = 1 TO P
0310 FOR K = 1 TO Q
0320 FOR I = 1 TO N
0330 READ R
0340 LET S1 = S1 + (R - M)*(R - M)
0350 LET M1 = M1 + R/(N*Q)
0360 LET M2 = M2 + R/N
0370 NEXT I
0380 GOSUB 3010
0390 IF L<> 0 THEN GOSUB 3050
0400 FOR I = 1 TO N
0410 READ R
0420 LET S5 = S5 + (R - M2)*(R - M2)
0430 LET V5 = V5 + (R - M2)*(R - M2)/(N - 1)
0440 NEXT I
0450 PRINT "MEAN A ";J;" B ";K;" = ";M2;"VAR. A ";J;" B ";K;" = ";V5
0460 GOSUB 3100
0470 LET S4 = S4 + (M2 - M)*(M2 - M)*N
0480 LET M2 = 0
0490 LET V5 = 0
0500 LET L = L + N
0510 GOSUB 3010
0520 GOSUB 3050
0530 NEXT K
0540 LET S2 = S2 + (M1 - M)*(M1 - M)*N*Q
0550 PRINT "MEAN LEVEL A ";J;" = ";M1
0560 GOSUB 3100
0570 LET M1 = 0
0580 LET V2 = 0
0590 NEXT J
0600 REM CALCULATE MEANS FOR LEVELS OF FACTOR B (M3),
0610 REM SSB (S3), AND COMPLETED VALUE FOR SSAB (S4)
0620 LET L = 0
0630 LET M3 = 0
0640 LET S3 = 0
0650 FOR K = 1 TO Q
0660 GOSUB 3010
0670 IF L = 0 THEN 690
0680 GOSUB 3050
0690 FOR J = 1 TO P
0700 FOR I = 1 TO N
0710 READ R
0720 LET M3 = M3 + R/(N*P)
0730 NEXT I
```

```
0740 GOSUB 3180
0750 NEXT J
0760 PRINT "MEAN B ";K;" = ";M3
0770 GOSUB 3110
0780 LET S3 = S3 + (M3 - M)*(M3 - M)*N*P
0790 LET L = L + N
0800 LET M3 = 0
0810 NEXT K
0820 LET S4 = S4 - S2 - S3
0840 REM CALCULATE MEAN FOR EACH LEVEL OF A FOR A GIVEN
0850 REM BLOCK (M6), SSS.W.GRP. (S6), AND SSBXS.W.GRP. (S7).
0860 LET M6 = 0
0870 LET S6 = 0
0880 LET G = 0
0890 LET H = 0
0900 FOR J = 1 TO P
0910 FOR I = 1 TO N
0920 GOSUB 3010
0930 IF H = 0 THEN 960
0940 LET L = H
0950 GOSUB 3050
0960 IF G = 0 THEN 990
0970 LET L = G
0980 GOSUB 3050
0990 FOR K = 1 TO Q
1000 READ R
1010 LET M6 = M6 + R/Q
1020 LET L = N - 1
1030 GOSUB 3050
1040 NEXT K
1050 LET S6 = S6 + (M6 - M)*(M6 - M)*Q - S2/(N*P)
1060 PRINT "MEAN LEVEL A";J;"N";I;" = ";M6
1070 GOSUB 3110
1080 LET M6 = 0
1090 LET G = G + 1
1100 NEXT I
1110 LET G = 0
1120 LET H = H + N*Q
1130 NEXT J
1140 LET S7 = S5 - S6
1160 REM CALCULATE MEAN BLOCK (M8), SSBLOCKS (S8),
1170 REM AND SSAXBLOCKS (S9).
1180 LET G = 0
1190 LET S8 = 0
1200 FOR I = 1 TO N
1210 LET M8 = 0
1220 GOSUB 3010
1230 IF G = 0 THEN 1260
1240 LET L = G
1250 GOSUB 3050
1260 FOR J = 1 TO P
1270 FOR K = 1 TO Q
1280 READ R
1290 LET M8 = M8 + R/(P*Q)
1300 LET L = N - 1
1310 GOSUB 3050
1320 NEXT K
1330 NEXT J
1340 LET S8 = S8 + (M8 - M)*(M8 - M)*P*Q
1350 PRINT "MEAN BLOCK ";I;" = ";M8
1360 GOSUB 3110
1370 LET G = G + 1
1380 NEXT I
1390 LET S9 = S6 - S8
1400 REM CALCULATE MEAN FOR EACH LEVEL OF B FOR A GIVEN
1410 REM BLOCK (M0), SSBXBLOCKS (S0), AND SSABXBLOCKS (T1)
1420 LET H = 0
1430 LET S0 = 0
1440 LET M0 = 0
1450 FOR K = 1 TO Q
1460 FOR I = 1 TO N
1470 GOSUB 3010
```

Program 11. (cont.)

```
1480 IF H = 0 THEN 1510
1490 LET L = H
1500 GOSUB 3050
1510 FOR J = 1 TO P
1520 READ R
1530 LET M0 = M0 + R/P
1540 LET L = N*Q - 1
1550 GOSUB 3050
1560 NEXT J
1570 LET S0 = S0 + (M0 - M)*(M0 - M)*P - S3/(Q*N) - S8/(Q*N)
1580 PRINT "MEAN LEVEL B";K;"N";I;" = ";M0
1590 GOSUB 3110
1600 LET H = H + 1
1610 LET M0 = 0
1620 NEXT I
1630 NEXT K
1640 LET T1 = S5 - S8 - S9 - S0
1660 LET D8 = N - 1
1670 LET D9 = D8*(P - 1)
1680 LET D0 = D8*(Q - 1)
1690 LET E1 = D9*(Q - 1)
1700 LET T = 13
1710 GOSUB 3110
1720 LET T2 = S2 + S3 + S4 + S5
1730 LET D1 = P*Q*N - 1
1740 LET D2 = P - 1
1750 LET D3 = Q - 1
1760 LET D4 = D2*D3
1770 LET D5 = P*Q*(N - 1)
1780 PRINT
1790 PRINT TAB(9);"SS";TAB(22);"DF";TAB(35);"MS";TAB(48);"F"
1800 PRINT "TOTAL";TAB(9);S1;TAB(22);D1
1810 PRINT "  BLOCK";TAB(9);S8;TAB(22);D8;TAB(35);S8/D8;TAB(48);
1820 PRINT (S8/D8)/((S9 + S0 + T1)/(D9 + D0 + E1))
1830 PRINT "  A";TAB(9);S2;TAB(22);D2;TAB(35);S2/D2;TAB(48);
1840 PRINT (S2/D2)/(S9/D9)
1850 PRINT "  AXBLK";TAB(9);S9;TAB(22);D9;TAB(35);S9/D9
1860 PRINT "  B";TAB(9);S3;TAB(22);D3;TAB(35);S3/D3;TAB(48);
1870 PRINT (S3/D3)/(S0/D0)
1880 PRINT "  BXBLK";TAB(9);S0;TAB(22);D0;TAB(35);S0/D0
1890 PRINT "  AB";TAB(9);S4;TAB(22);D4;TAB(35);S4/D4;TAB(48);
1900 PRINT (S4/D4)/(T1/E1)
1910 PRINT "  ABXBK";TAB(9);T1;TAB(22);E1;TAB(35);T1/E1
1920 PRINT "TOTAL";TAB(9);S2 + S3 + S4 + S5;TAB(22);D1
1930 GOTO 9999
3000 REM RESTORE DATA POINTER AND ADVANCE DATA POINTER
3010 RESTORE
3020 READ R, R, R, R
3030 RETURN
3040 REM ADVANCE DATA POINTER
3050 FOR L1 = 1 TO L
3060 READ R
3070 IF R > 2E22 THEN 3090
3080 NEXT L1
3090 RETURN
3100 REM SUBROUTINE TO STOP EXECUTION UNTIL DATA ARE TRANSCRIBED
3110 LET T = T + 1
3120 IF T <= 12 THEN RETURN
3130 PRINT "ENTER 1 FOLLOWED BY CARRIAGE RETURN TO CONTINUE."
3140 INPUT R1
3150 LET T = 0
3160 RETURN
3170 REM SUBROUTINE TO ADVANCE DATA POINTER TO NEXT LEVEL OF FACTOR B
3180 FOR L1 = 1 TO Q - 1
3190 FOR I = 1 TO N
3200 READ R
3210 IF R > 2E22 THEN 3240
3220 NEXT I
3230 NEXT L1
3240 RETURN
9998 DATA 2E23
9999 END
```

Program 12. Completely randomized factorial analysis of variance (requires equal n in all cells), CRF-JKL.

```
0010 REM THIS PROGRAM IS FOR AN INDEPENDENT GROUPS THREE-
0020 REM WAY ANALYSIS OF VARIANCE (COMPLETELY RANDOMIZED
0030 REM FACTORIAL ANALYSIS OF VARIANCE (CRF-JKL)).
0080 PRINT
0090 PRINT
0100 REM SEE APPROPRIATE EXAMPLE FOR ENTERING DATA
0110 READ A, N, P, Q, R
0120 REM CALCULATE GRAND MEAN
0130 GOSUB 3010
0140 LET M = 0
0150 FOR I = 1 TO N*P*Q*R
0160 READ S
0170 LET M = M + S/(N*P*Q*R)
0180 NEXT I
0190 REM CALCULATE MEAN EACH LEVEL FACTOR B (M3), AND SSB (S3)
0200 LET S3 = 0
0210 LET M3 = 0
0220 LET G = 0
0230 GOSUB 3010
0240 FOR K = 1 TO Q
0250 FOR J = 1 TO P
0260 FOR L = 1 TO R
0270 FOR I = 1 TO N
0280 READ S
0290 LET M3 = M3 + S/(N*P*R)
0300 NEXT I
0310 NEXT L
0320 LET H = N*R*(Q - 1)
0330 GOSUB 3050
0340 NEXT J
0350 LET S3 = S3 + (M3 - M)*(M3 - M)*N*P*R
0360 PRINT "MEAN B";K;" = ";M3
0370 GOSUB 3110
0380 LET M3 = 0
0390 LET G = G + N*R
0400 LET H = G
0410 GOSUB 3010
0420 GOSUB 3050
0430 NEXT K
0440 REM CALCULATE SSTOTAL (S1), MEAN EACH LEVEL
0450 REM FACTOR A (M2), SSA (S2), MEAN EACH LEVEL
0460 REM AB INTERACTION (M5), SSAB (S5), MEAN EACH LEVEL
0470 REM ABC INTERACTION (M8), VAR. EACH LEVEL ABC (V9),
0480 REM AND SSWITHIN CELLS (S9)
0490 LET S1 = 0
0500 LET S2 = 0
0510 LET S5 = 0
0520 LET S9 = 0
0530 LET M2 = 0
0540 LET M5 = 0
0550 LET M8 = 0
0560 LET V9 = 0
0570 LET G1 = 0
0580 LET G2 = 0
0590 FOR J = 1 TO P
0600 FOR K = 1 TO Q
0610 FOR L = 1 TO R
0620 GOSUB 3010
0630 IF G2 = 0 THEN 660
0640 LET H = G2
0650 GOSUB 3050
0660 IF G1 <> 0 THEN 750
0670 FOR I = 1 TO N
0680 READ S
0690 LET M2 = M2 + S/(N*Q*R)
0700 LET M5 = M5 + S/(N*R)
0710 LET M8 = M8 + S/N
0720 NEXT I
0730 LET G1 = 1
0740 GOTO 620
```

Program 12. (cont.)

```
0750 FOR I = 1 TO N
0760 READ S
0770 LET S1 = S1 + (S - M)*(S - M)
0780 LET S9 = S9 + (S - M8)*(S - M8)
0790 LET V9 = V9 + (S - M8)*(S - M8)/(N - 1)
0800 NEXT I
0810 PRINT "MEAN CELL A";J;"B";K;"C";L;" = ";M8;
0820 PRINT "VAR. CELL A";J;"B";K;"C";L;" = ";V9
0830 LET V9 = 0
0840 GOSUB 3110
0850 LET M8 = 0
0860 LET G1 = 0
0870 LET G2 = G2 + N
0880 NEXT L
0890 LET S5 = S5 + (M5 - M)*(M5 - 5)*N*R - S3/(P*Q)
0900 PRINT "MEAN A";J;"B";K;" = ";M5
0910 GOSUB 3110
0920 LET M5 = 0
0930 NEXT K
0940 LET S2 = S2 + (M2 - M)*(M2 - M)*N*Q*R
0950 PRINT "MEAN A";J;" = ";M2
0960 GOSUB 3110
0970 LET M2 = 0
0980 NEXT J
0990 LET S5 = S5 - S2
1000 REM CALCULATE MEAN EACH LEVEL FACTOR C (M4), AND SSC (S4)
1010 LET G = 0
1020 LET S4 = 0
1030 LET M4 = 0
1040 GOSUB 3010
1050 FOR L = 1 TO R
1060 FOR J = 1 TO P
1070 FOR K = 1 TO Q
1080 FOR I = 1 TO N
1090 READ S
1100 LET M4 = M4 + S/(N*P*Q)
1110 NEXT I
1120 LET H = N*(R - 1)
1130 GOSUB 3050
1140 NEXT K
1150 NEXT J
1160 LET S4 = S4 + (M4 - M)*(M4 - M)*N*P*Q
1170 PRINT "MEAN C";L;" = ";M4
1180 GOSUB 3110
1190 LET M4 = 0
1200 GOSUB 3010
1210 LET G = G + N
1220 LET H = G
1230 GOSUB 3050
1240 NEXT L
1250 REM CALCULATE MEAN EACH LEVEL AC (M6), AND SSAC (S6)
1260 LET M6 = 0
1270 LET S6 = 0
1280 LET G = 0
1290 LET G1 = 0
1300 GOSUB 3010
1310 FOR J = 1 TO P
1320 FOR L = 1 TO R
1330 FOR K = 1 TO Q
1340 FOR I = 1 TO N
1350 READ S
1360 LET M6 = M6 + S/(N*Q)
1370 NEXT I
1380 LET H = N*R - N
1390 GOSUB 3050
1400 NEXT K
1410 LET S6 = S6 + (M6 - M)*(M6 - M)*N*Q - S2/(P*R) - S4/(P*R)
1420 PRINT "MEAN A";J;"C";L;" = ";M6
1430 GOSUB 3110
1440 LET M6 = 0
1450 LET G = G + N
1460 GOSUB 3010
```

Program 12. (cont.)

```
1470 LET H = G
1480 GOSUB 3050
1490 IF G1 = 0 THEN 1520
1500 LET H = G1
1510 GOSUB 3050
1520 NEXT L
1530 LET G = 0
1540 LET G1 = G1 + N*Q*R
1550 GOSUB 3010
1560 LET H = G1
1570 GOSUB 3050
1580 NEXT J
1590 REM CALCULATE MEAN EACH LEVEL BC (M7), SSBC (S7) AND SSABC (S8)
1600 LET S7 = 0
1610 LET S8 = 0
1620 LET M7 = 0
1630 LET M8 = 0
1640 LET G = 0
1650 GOSUB 3010
1660 FOR K = 1 TO Q
1670 FOR L = 1 TO R
1680 FOR J = 1 TO P
1690 FOR I = 1 TO N
1700 READ S
1710 LET M7 = M7 + S/(N*P)
1720 LET M8 = M8 + S/N
1730 NEXT I
1740 LET G1 = P*Q*R
1750 LET S8 = S8 + (M8 - M)*(M8 - M)*N
1760 LET S8 = S8 - S2/G1 - S3/G1 - S4/G1 - S5/G1 - S6/G1
1770 LET M8 = 0
1780 LET H = N*R*Q - N
1790 GOSUB 3050
1800 NEXT J
1810 PRINT "MEAN B";K;"C";L;" = ";M7
1820 GOSUB 3110
1830 LET S7 = S7 + (M7 - M)*(M7 - M)*N*P - S3/(Q*R) - S4/(Q*R)
1840 LET M7 = 0
1850 LET G = G + N
1860 GOSUB 3010
1870 LET H = G
1880 GOSUB 3050
1890 NEXT L
1900 NEXT K
2000 LET S8 = S8 - S7
2010 LET D1 = N*P*Q*R - 1
2020 LET D2 = P - 1
2030 LET D3 = Q - 1
2040 LET D4 = R - 1
2050 LET D5 = D2*D3
2060 LET D6 = D2*D4
2070 LET D7 = D3*D4
2080 LET D8 = D5*D4
2090 LET D9 = P*Q*R*(N - 1)
2100 LET T = 13
2110 GOSUB 3120
2120 PRINT
2130 PRINT TAB(9);"SS";TAB(22);"DF";TAB(35);"MS";TAB(48);"F"
2140 PRINT "TOTAL";TAB(9);S1;TAB(22);D1
2150 PRINT "  A";TAB(9);S2;TAB(22);D2;TAB(35);S2/D2;TAB(48);
2160 PRINT (S2/D2)/(S9/D9)
2170 PRINT "  B";TAB(9);S3;TAB(22);D3;TAB(35);S3/D3;TAB(48);
2180 PRINT (S3/D3)/(S9/D9)
2190 PRINT "  C";TAB(9);S4;TAB(22);D4;TAB(35);S4/D4;TAB(48);
2200 PRINT (S4/D4)/(S9/D9)
2210 PRINT "  AB";TAB(9);S5;TAB(22);D5;TAB(35);S5/D5;TAB(48);
2220 PRINT (S5/D5)/(S9/D9)
2230 PRINT "  AC";TAB(9);S6;TAB(22);D6;TAB(35);S6/D6;TAB(48);
2240 PRINT (S6/D6)/(S9/D9)
2250 PRINT "  BC";TAB(9);S7;TAB(22);D7;TAB(35);S7/D7;TAB(48);
2260 PRINT (S7/D7)/(S9/D9)
2270 PRINT "  ABC";TAB(9);S8;TAB(22);D8;TAB(35);S8/D8;TAB(48);
```

Program 12. (cont.)

```
2280 PRINT (S8/D8)/(S9/D9)
2290 PRINT "  W.CEL";TAB(9);S9;TAB(22);D9;TAB(35);S9/D9
2300 PRINT "TOTAL";TAB(9);S2+S3+S4+S5+S6+S7+S8+S9;TAB(22);D1
2310 GOTO 9999
3000 REM RESTORE DATA POINTER AND ADVANCE DATA POINTER
3010 RESTORE
3020 READ S, S, S, S, S
3030 RETURN
3040 REM ADVANCE DATA POINTER
3050 FOR L1 = 1 TO H
3060 READ S
3070 IF S > 2E22 THEN 3090
3080 NEXT L1
3090 RETURN
3100 REM SUBROUTINE TO STOP EXECUTION UNTIL DATA ARE TRANSCRIBED
3110 LET T = T + 1
3120 IF T <= 10 THEN RETURN
3130 PRINT "ENTER 1 FOLLOWED BY CARRIAGE RETURN TO CONTINUE."
3140 INPUT R1
3150 LET T = 0
3160 RETURN
9998 DATA 2E23
9999 END
```

Program 13. Split-plot analysis of variance (requires equal n in all cells), SPF-JK.L.

```
0010 REM THIS PROGRAM IS FOR A SPLIT-PLOT ANALYSIS
0020 REM OF VARIANCE (ONE REPEATED MEASURE, SPF-JK.L)
0080 PRINT
0090 PRINT
0100 REM SEE APPROPRIATE EXAMPLE FOR ENTERING DATA
0110 READ A, N, P, Q, R
0120 REM CALCULATE GRAND MEAN
0130 GOSUB 3010
0140 LET M = 0
0150 FOR I = 1 TO N*P*Q*R
0160 READ S
0170 LET M = M + S/(N*P*Q*R)
0180 NEXT I
0190 REM CALCULATE MEAN EACH LEVEL FACTOR B (M3), AND SSB (S3)
0200 LET S3 = 0
0210 LET M3 = 0
0220 LET G = 0
0230 GOSUB 3010
0240 FOR K = 1 TO Q
0250 FOR J = 1 TO P
0260 FOR L = 1 TO R
0270 FOR I = 1 TO N
0280 READ S
0290 LET M3 = M3 + S/(N*P*R)
0300 NEXT I
0310 NEXT L
0320 LET H = N*R*(Q - 1)
0330 GOSUB 3050
0340 NEXT J
0350 LET S3 = S3 + (M3 - M)*(M3 - M)*N*P*R
0360 PRINT "MEAN B";K;" = ";M3
0370 GOSUB 3110
0380 LET M3 = 0
0390 LET G = G + N*R
0400 LET H = G
0410 GOSUB 3010
0420 GOSUB 3050
0430 NEXT K
0440 REM CALCULATE SSTOTAL (S1), MEAN EACH LEVEL
0450 REM FACTOR A (M2), SSA (S2), MEAN EACH LEVEL
0460 REM AB INTERACTINON (M5), SSAB (S5), MEAN EACH LEVEL
0470 REM ABC INTERACTION (M8), VAR. EACH LEVEL ABC (V9)
0480 REM AND SSWITHIN CELLS (S9)
0490 LET S1 = 0
0500 LET S2 = 0
0510 LET S5 = 0
0520 LET S9 = 0
0530 LET M2 = 0
0540 LET M5 = 0
0550 LET M8 = 0
0560 LET V9 = 0
0570 LET G1 = 0
0580 LET G2 = 0
0590 FOR J = 1 TO P
0600 FOR K = 1 TO Q
0610 FOR L = 1 TO R
0620 GOSUB 3010
0630 IF G2 = 0 THEN 660
0640 LET H = G2
0650 GOSUB 3050
0660 IF G1 <> 0 THEN 750
0670 FOR I = 1 TO N
0680 READ S
0690 LET M2 = M2 + S/(N*Q*R)
0700 LET M5 = M5 + S/(N*R)
0710 LET M8 = M8 + S/N
0720 NEXT I
0730 LET G1 = 1
0740 GOTO 620
0750 FOR I = 1 TO N
```

Program 13. (cont.)

```
0760 READ S
0770 LET S1 = S1 + (S - M)*(S - M)
0780 LET S9 = S9 + (S - M8)*(S - M8)
0790 LET V9 = V9 + (S - M8)*(S - M8)/(N - 1)
0800 NEXT I
0810 PRINT "MEAN CELL A";J;"B";K;"C";L;" = ";M8;
0820 PRINT "VAR. CELL A";J;"B";K;"C";L;" = ";V9
0830 LET V9 = 0
0840 GOSUB 3110
0850 LET M8 = 0
0860 LET G1 = 0
0870 LET G2 = G2 + N
0880 NEXT L
0890 LET S5 = S5 + (M5 - M)*(M5 - 5)*N*R - S3/(P*Q)
0900 PRINT "MEAN A";J;"B";K;" = ";M5
0910 GOSUB 3110
0920 LET M5 = 0
0930 NEXT K
0940 LET S2 = S2 + (M2 - M)*(M2 - M)*N*Q*R
0950 PRINT "MEAN A";J;" = ";M2
0960 GOSUB 3110
0970 LET M2 = 0
0980 NEXT J
0990 LET S5 = S5 - S2
1000 REM CALC. MEAN EACH LEVEL FACTOR C (M4), AND SSC (S4)
1010 LET G = 0
1020 LET S4 = 0
1030 LET M4 = 0
1040 GOSUB 3010
1050 FOR L = 1 TO R
1060 FOR J = 1 TO P
1070 FOR K = 1 TO Q
1080 FOR I = 1 TO N
1090 READ S
1100 LET M4 = M4 + S/(N*P*Q)
1110 NEXT I
1120 LET H = N*(R - 1)
1130 GOSUB 3050
1140 NEXT K
1150 NEXT J
1160 LET S4 = S4 + (M4 - M)*(M4 - M)*N*P*Q
1170 PRINT "MEAN C";L;" = ";M4
1180 GOSUB 3110
1190 LET M4 = 0
1200 GOSUB 3010
1210 LET G = G + N
1220 LET H = G
1230 GOSUB 3050
1240 NEXT L
1250 REM CALCULATE MEAN EACH LEVEL AC (M6), AND SSAC (S6)
1260 LET M6 = 0
1270 LET S6 = 0
1280 LET G = 0
1290 LET G1 = 0
1300 GOSUB 3010
1310 FOR J = 1 TO P
1320 FOR L = 1 TO R
1330 FOR K = 1 TO Q
1340 FOR I = 1 TO N
1350 READ S
1360 LET M6 = M6 + S/(N*Q)
1370 NEXT I
1380 LET H = N*R - N
1390 GOSUB 3050
1400 NEXT K
1410 LET S6 = S6 + (M6 - M)*(M6 - M)*N*Q - S2/(P*R) - S4/(P*R)
1420 PRINT "MEAN A";J;"C";L;" = ";M6
1430 GOSUB 3110
1440 LET M6 = 0
1450 LET G = G + N
```

```
1460 GOSUB 3010
1470 LET H = G
1480 GOSUB 3050
1490 IF G1 = 0 THEN 1520
1500 LET H = G1
1510 GOSUB 3050
1520 NEXT L
1530 LET G = 0
1540 LET G1 = G1 + N*Q*R
1550 GOSUB 3010
1560 LET H = G1
1570 GOSUB 3050
1580 NEXT J
1590 REM CALC. MEAN EACH LEVEL BC(M7), SSBC(S7), AND SSABC(S8)
1600 LET S7 = 0
1610 LET S8 = 0
1620 LET M7 = 0
1630 LET M8 = 0
1640 LET G = 0
1650 GOSUB 3010
1660 FOR K = 1 TO Q
1670 FOR L = 1 TO R
1680 FOR J = 1 TO P
1690 FOR I = 1 TO N
1700 READ S
1710 LET M7 = M7 + S/(N*P)
1720 LET M8 = M8 + S/N
1730 NEXT I
1740 LET G1 = P*Q*R
1750 LET S8 = S8 + (M8 - M)*(M8 - M)*N
1760 LET S8 = S8 - S2/G1 - S3/G1 - S4/G1 - S5/G1 - S6/G1
1770 LET M8 = 0
1780 LET H = N*R*Q - N
1790 GOSUB 3050
1800 NEXT J
1810 PRINT "MEAN B";K;"C";L;" = ";M7
1820 GOSUB 3110
1830 LET S7 = S7 + (M7 - M)*(M7 - M)*N*P - S3/(Q*R) - S4/(Q*R)
1840 LET M7 = 0
1850 LET G = G + N
1860 GOSUB 3010
1870 LET H = G
1880 GOSUB 3050
1890 NEXT L
1900 NEXT K
2000 LET S8 = S8 - S7
2010 LET D1 = N*P*Q*R - 1
2020 LET D2 = P - 1
2030 LET D3 = Q - 1
2040 LET D4 = R - 1
2050 LET D5 = D2*D3
2060 LET D6 = D2*D4
2070 LET D7 = D3*D4
2080 LET D8 = D5*D4
2090 LET D9 = P*Q*R*(N - 1)
2100 REM CALCULATE SSS.W.GP (T1), SSCXS.W.GP (T2)
2110 LET M1 = 0
2120 LET T1 = 0
2130 LET U1 = 0
2140 LET G = 0
2150 GOSUB 3010
2160 FOR I = 1 TO N
2170 FOR J = 1 TO P
2180 FOR K = 1 TO Q
2190 FOR L = 1 TO R
2200 READ S
2210 LET M1 = M1 + S/R
2220 LET H = N - 1
2230 GOSUB 3050
2240 NEXT L
2250 LET G1 = N*P*Q
```

Program 13. (cont.)

```
2260 LET T1 = T1 + (M1 - M)*(M1 - M)*R -S2/G1 -S3/G1 -S5/G1
2270 LET M1 = 0
2280 NEXT K
2290 NEXT J
2300 GOSUB 3010
2310 LET G = G + 1
2320 LET H = G
2330 GOSUB 3050
2340 NEXT I
2350 LET T2 = S9 - T1
2360 LET E1 = P*Q*(N - 1)
2370 LET E2 = P*Q*(R - 1)*(N - 1)
2380 LET T = 13
2390 GOSUB 3120
2400 PRINT
2410 PRINT TAB(9);"SS";TAB(22);"DF";TAB(35);"MS";TAB(48);"F"
2420 PRINT "TOTAL";TAB(9);S1;TAB(22);D1
2430 PRINT " BET.";TAB(9);S2+S3+S5+T1;TAB(22);D2+D3+D5+E1
2440 PRINT "  A";TAB(9);S2;TAB(22);D2;TAB(35);S2/D2;TAB(48);
2450 PRINT (S2/D2)/(T1/E1)
2460 PRINT "  B";TAB(9);S3;TAB(22);D3;TAB(35);S3/D3;TAB(48);
2470 PRINT (S3/D3)/(T1/E1)
2480 PRINT "  AB";TAB(9);S5;TAB(22);D5;TAB(35);S5/D5;TAB(48);
2490 PRINT (S5/D5)/(T1/E1)
2500 PRINT "  SWG.";TAB(9);T1;TAB(22);E1;TAB(35);T1/E1
2510 PRINT " WITH";TAB(9);S4 + S6 + S7 + S8 + T2;TAB(22);
2520 PRINT D4 + D6 + D7 + D8 + E2
2530 PRINT "  C";TAB(9);S4;TAB(22);D4;TAB(35);S4/D4;TAB(48);
2540 PRINT (S4/D4)/(T2/E2)
2550 PRINT "  AC";TAB(9);S6;TAB(22);D6;TAB(35);S6/D6;TAB(48);
2560 PRINT (S6/D6)/(T2/E2)
2570 PRINT "  BC";TAB(9);S7;TAB(22);D7;TAB(35);S7/D7;TAB(48);
2580 PRINT (S7/D7)/(T2/E2)
2590 PRINT "  ABC";TAB(9);S8;TAB(22);D8;TAB(35);S8/D8;TAB(48);
2600 PRINT (S8/D8)/(T2/E2)
2610 PRINT "  CXSWG";TAB(9);T2;TAB(22);E2;TAB(35);T2/E2
2620 PRINT "TOTAL";TAB(9);S2+S3+S4+S5+S6+S7+S8+S9;TAB(22);D1
2630 GOTO 9999
3000 REM RESTORE DATA POINTER AND ADVANCE DATA POINTER
3010 RESTORE
3020 READ S, S, S, S, S
3030 RETURN
3040 REM ADVANCE DATA POINTER
3050 FOR L1 = 1 TO H
3060 READ S
3070 IF S > 2E22 THEN 3090
3080 NEXT L1
3090 RETURN
3100 REM SUBROUTINE TO STOP EXECUTION UNTIL DATA ARE TRANSCRIBED
3110 LET T = T + 1
3120 IF T <= 10 THEN RETURN
3130 PRINT "ENTER 1 FOLLOWED BY CARRIAGE RETURN TO CONTINUE."
3140 INPUT R1
3150 LET T = 0
3160 RETURN
9998 DATA 2E23
9999 END
```

Program 14. Split-plot analysis of variance (requires equal n in all cells), SPF-J.KL.

```
0010 REM THIS PROGRAM IS FOR A SPLIT-PLOT ANALYSIS
0020 REM OF VARIANCE (TWO REPEATED MEASURES, SPF-J.KL)
0080 PRINT
0090 PRINT
0100 REM SEE APPROPRIATE EXAMPLE FOR ENTERING DATA
0110 READ A, N, P, Q, R
0120 REM CALCULATE GRAND MEAN
0130 GOSUB 3010
0140 LET M = 0
0150 FOR I = 1 TO N*P*Q*R
0160 READ S
0170 LET M = M + S/(N*P*Q*R)
0180 NEXT I
0190 REM CALCULATE MEAN EACH LEVEL FACTOR B (M3), AND SSB (S3)
0200 LET S3 = 0
0210 LET M3 = 0
0220 LET G = 0
0230 GOSUB 3010
0240 FOR K = 1 TO Q
0250 FOR J = 1 TO P
0260 FOR L = 1 TO R
0270 FOR I = 1 TO N
0280 READ S
0290 LET M3 = M3 + S/(N*P*R)
0300 NEXT I
0310 NEXT L
0320 LET H = N*R*(Q - 1)
0330 GOSUB 3050
0340 NEXT J
0350 LET S3 = S3 + (M3 - M)*(M3 - M)*N*P*R
0360 PRINT "MEAN B";K;" = ";M3
0370 GOSUB 3110
0380 LET M3 = 0
0390 LET G = G + N*R
0400 LET H = G
0410 GOSUB 3010
0420 GOSUB 3050
0430 NEXT K
0440 REM CALCULATE SSTOTAL (S1), MEAN EACH LEVEL
0450 REM FACTOR A (M2), SSA (S2), MEAN EACH LEVEL
0460 REM AB INTERACTION (M5), SSAB (S5), MEAN EACH LEVEL
0470 REM ABC INTERACTION (M8), VAR. EACH LEVEL ABC (V9)
0480 REM AND SSWITHIN CELLS (S9)
0490 LET S1 = 0
0500 LET S2 = 0
0510 LET S5 = 0
0520 LET S9 = 0
0530 LET M2 = 0
0540 LET M5 = 0
0550 LET M8 = 0
0560 LET V9 = 0
0570 LET G1 = 0
0580 LET G2 = 0
0590 FOR J = 1 TO P
0600 FOR K = 1 TO Q
0610 FOR L = 1 TO R
0620 GOSUB 3010
0630 IF G2 = 0 THEN 660
0640 LET H = G2
0650 GOSUB 3050
0660 IF G1 <> 0 THEN 750
0670 FOR I = 1 TO N
0680 READ S
0690 LET M2 = M2 + S/(N*Q*R)
0700 LET M5 = M5 + S/(N*R)
0710 LET M8 = M8 + S/N
0720 NEXT I
0730 LET G1 = 1
0740 GOTO 620
0750 FOR I = 1 TO N
```

Program 14. (cont.)

```
0760 READ S
0770 LET S1 = S1 + (S - M)*(S - M)
0780 LET S9 = S9 + (S - M8)*(S - M8)
0790 LET V9 = V9 + (S - M8)*(S - M8)/(N - 1)
0800 NEXT I
0810 PRINT "MEAN CELL A";J;"B";K;"C";L;" = ";M8;
0820 PRINT "VAR. CELL A";J;"B";K;"C";L;" = ";V9
0830 LET V9 = 0
0840 GOSUB 3110
0850 LET M8 = 0
0860 LET G1 = 0
0870 LET G2 = G2 + N
0880 NEXT L
0890 LET S5 = S5 + (M5 - M)*(M5 - 5)*N*R - S3/(P*Q)
0900 PRINT "MEAN A";J;"B";K;" = ";M5
0910 GOSUB 3110
0920 LET M5 = 0
0930 NEXT K
0940 LET S2 = S2 + (M2 - M)*(M2 - M)*N*Q*R
0950 PRINT "MEAN A";J;" = ";M2
0960 GOSUB 3110
0970 LET M2 = 0
0980 NEXT J
0990 LET S5 = S5 - S2
1000 REM CALCULATE MEAN EACH LEVEL FACTOR C (M4), AND SSC (S4)
1010 LET G = 0
1020 LET S4 = 0
1030 LET M4 = 0
1040 GOSUB 3010
1050 FOR L = 1 TO R
1060 FOR J = 1 TO P
1070 FOR K = 1 TO Q
1080 FOR I = 1 TO N
1090 READ S
1100 LET M4 = M4 + S/(N*P*Q)
1110 NEXT I
1120 LET H = N*(R - 1)
1130 GOSUB 3050
1140 NEXT K
1150 NEXT J
1160 LET S4 = S4 + (M4 - M)*(M4 - M)*N*P*Q
1170 PRINT "MEAN C";L;" = ";M4
1180 GOSUB 3110
1190 LET M4 = 0
1200 GOSUB 3010
1210 LET G = G + N
1220 LET H = G
1230 GOSUB 3050
1240 NEXT L
1250 REM CALCULATE MEAN EACH LEVEL AC (M6), AND SSAC (S6)
1260 LET M6 = 0
1270 LET S6 = 0
1280 LET G = 0
1290 LET G1 = 0
1300 GOSUB 3010
1310 FOR J = 1 TO P
1320 FOR L = 1 TO R
1330 FOR K = 1 TO Q
1340 FOR I = 1 TO N
1350 READ S
1360 LET M6 = M6 + S/(N*Q)
1370 NEXT I
1380 LET H = N*R - N
1390 GOSUB 3050
1400 NEXT K
1410 LET S6 = S6 + (M6 - M)*(M6 - M)*N*Q - S2/(P*R) - S4/(P*R)
1420 PRINT "MEAN A";J;"C";L;" = ";M6
1430 GOSUB 3110
1440 LET M6 = 0
1450 LET G = G + N
1460 GOSUB 3010
1470 LET H = G
```

```
1480 GOSUB 3050
1490 IF G1 = 0 THEN 1520
1500 LET H = G1
1510 GOSUB 3050
1520 NEXT L
1530 LET G = 0
1540 LET G1 = G1 + N*Q*R
1550 GOSUB 3010
1560 LET H = G1
1570 GOSUB 3050
1580 NEXT J
1590 REM CALCULATE MEAN EACH LEVEL BC (M7), SSBC (S7) AND SSABC (S8)
1600 LET S7 = 0
1610 LET S8 = 0
1620 LET M7 = 0
1630 LET M8 = 0
1640 LET G = 0
1650 GOSUB 3010
1660 FOR K = 1 TO Q
1670 FOR L = 1 TO R
1680 FOR J = 1 TO P
1690 FOR I = 1 TO N
1700 READ S
1710 LET M7 = M7 + S/(N*P)
1720 LET M8 = M8 + S/N
1730 NEXT I
1740 LET G1 = P*Q*R
1750 LET S8 = S8 + (M8 - M)*(M8 - M)*N
1760 LET S8 = S8 - S2/G1 - S3/G1 - S4/G1 - S5/G1 - S6/G1
1770 LET M8 = 0
1780 LET H = N*R*Q - N
1790 GOSUB 3050
1800 NEXT J
1810 PRINT "MEAN B";K;"C";L;" = ";M7
1820 GOSUB 3110
1830 LET S7 = S7 + (M7 - M)*(M7 - M)*N*P - S3/(Q*R) - S4/(Q*R)
1840 LET M7 = 0
1850 LET G = G + N
1860 GOSUB 3010
1870 LET H = G
1880 GOSUB 3050
1890 NEXT L
1900 NEXT K
2000 LET S8 = S8 - S7
2010 LET D1 = N*P*Q*R - 1
2020 LET D2 = P - 1
2030 LET D3 = Q - 1
2040 LET D4 = R - 1
2050 LET D5 = D2*D3
2060 LET D6 = D2*D4
2070 LET D7 = D3*D4
2080 LET D8 = D5*D4
2090 LET D9 = P*Q*R*(N - 1)
2100 REM CALCULATE SSS.W.GP (T1), SSCXS.W.GP (T2)
2110 REM CALCULATE SSS.W.GP (T3),
2120 LET M1 = 0
2130 LET M3 = 0
2140 LET T1 = 0
2150 LET T3 = 0
2160 LET G = 0
2170 GOSUB 3010
2180 FOR I = 1 TO N
2190 FOR J = 1 TO P
2200 FOR K = 1 TO Q
2210 FOR L = 1 TO R
2220 READ S
2230 LET M1 = M1 + S/R
2240 LET M3 = M3 + S/(Q*R)
2250 LET H = N - 1
2260 GOSUB 3050
2270 NEXT L
```

Program 14. (cont.)

```
2280 LET G1 = N*P*Q
2290 LET T1 = T1 + (M1 - M)*(M1 - M)*R - S2/G1 - S3/G1 - S5/G1
2300 LET M1 = 0
2310 NEXT K
2320 LET T3 = T3 + (M3 - M)*(M3 - M)*Q*R - S2/(N*P)
2330 LET M3 = 0
2340 NEXT J
2350 GOSUB 3010
2360 LET G = G + 1
2370 LET H = G
2380 GOSUB 3050
2390 NEXT I
2400 LET T4 = T1 - T3
2410 REM CALCULATE SSCXS.W.GP (T5), AND SSBCXS.W.GP (T6)
2420 LET G = 0
2430 LET G1 = 0
2440 LET T5 = 0
2450 LET M5 = 0
2460 GOSUB 3010
2470 FOR L = 1 TO R
2480 FOR I = 1 TO N
2490 FOR J = 1 TO P
2500 FOR K = 1 TO Q
2510 READ S
2520 LET M5 = M5 + S/Q
2530 LET H = N*R - 1
2540 GOSUB 3050
2550 NEXT K
2560 LET G2 = N*P*R
2570 LET T5 = T5 + (M5 - M)*(M5 - M)*Q - S2/G2 - S4/G2 - S6/G2 - T3/G2
2580 LET M5 = 0
2590 NEXT J
2600 GOSUB 3010
2610 LET G = G + 1
2620 LET H = G
2630 GOSUB 3050
2640 NEXT I
2650 GOSUB 3010
2660 LET H = G
2670 GOSUB 3050
2680 NEXT L
2690 LET T6 = S9 - T3 - T4 - T5
2700 LET E3 = P*(N - 1)
2710 LET E4 = P*(Q - 1)*(N - 1)
2720 LET E5 = P*(R - 1)*(N - 1)
2730 LET E6 = E4*(R - 1)
2740 LET T = 13
2750 GOSUB 3120
2760 PRINT
2770 PRINT TAB(9);"SS";TAB(22);"DF";TAB(35);"MS";TAB(48);"F"
2780 PRINT "TOTAL";TAB(9);S1;TAB(22);D1
2790 PRINT " BET.";TAB(9);S2 + T3;TAB(22);D2 + E3
2800 PRINT "  A";TAB(9);S2;TAB(22);D2;TAB(35);S2/D2;TAB(48);S2/D2/T3*E3
2810 PRINT "  SWG.";TAB(9);T3;TAB(22);E3;TAB(35);T3/E3
2820 PRINT " WITH";TAB(9);S3 + S5 + T4 + S4 + S6 + T5 + S7 + S8 + T6;
2830 PRINT TAB(22);D3 + D5 + E4 + D4 + D6 + E5 + D7 + D8 + E6
2840 PRINT "  B";TAB(9);S3;TAB(22);D3;TAB(35);S3/D3;TAB(48);S3/D3/T4*E4
2850 PRINT "  AB";TAB(9);S5;TAB(22);D5;TAB(35);S5/D5;TAB(48);S5/D5/T4*E4
2860 PRINT "  BXSWG";TAB(9);T4;TAB(22);E4;TAB(35);T4/E4;TAB(48)
2870 PRINT "  C";TAB(9);S4;TAB(22);D4;TAB(35);S4/D4;TAB(48);S4/D4/T5*E5
2880 PRINT "  AC";TAB(9);S6;TAB(22);D6;TAB(35);S6/D6;TAB(48);S6/D6/T5*E5
2890 PRINT "  CXSWG";TAB(9);T5;TAB(22);E5;TAB(35);T5/E5
2900 PRINT "  BC";TAB(9);S7;TAB(22);D7;TAB(35);S7/D7;TAB(48);S7/D7/T6*E6
2910 PRINT "  ABC";TAB(9);S8;TAB(22);D8;TAB(35);S8/D8;TAB(48);
2920 PRINT S8/D8/T6*E6
2930 PRINT "  BCXSG";TAB(9);T6;TAB(22);E6;TAB(35);T6/E6
2940 PRINT "TOTAL";TAB(9);S2+S3+S4+S5+S6+S7+S8+S9;TAB(22);D1
2950 GOTO 9999
3000 REM RESTORE DATA POINTER AND ADVANCE DATA POINTER
3010 RESTORE
3020 READ S, S, S, S, S
```

Program 14. (cont.)

```
3030 RETURN
3040 REM ADVANCE DATA POINTER
3050 FOR L1 = 1 TO H
3060 READ S
3070 IF S > 2E22 THEN 3090
3080 NEXT L1
3090 RETURN
3100 REM SUBROUTINE TO STOP EXECUTION UNTIL DATA ARE TRANSCRIBED
3110 LET T = T + 1
3120 IF T <= 10 THEN RETURN
3130 PRINT "ENTER 1 FOLLOWED BY CARRIAGE RETURN TO CONTINUE."
3140 INPUT R1
3150 LET T = 0
3160 RETURN
9998 DATA 2E23
9999 END
```

Program 15. Randomized block factorial analysis of variance (requires equal n in all cells), RBF-.JKL.

```
0010 REM THIS PROGRAM IS FOR A RANDOMIZED BLOCK
0020 REM FACTORIAL ANALYSIS OF VARIANCE (RBF-.JKL)
0080 PRINT
0090 PRINT
0100 REM SEE APPROPRIATE EXAMPLE FOR ENTERING DATA
0110 READ A, N, P, Q, R
0120 REM CALCULATE GRAND MEAN
0130 GOSUB 3010
0140 LET M = 0
0150 FOR I = 1 TO N*P*Q*R
0160 READ S
0170 LET M = M + S/(N*P*Q*R)
0180 NEXT I
0190 REM CALCULATE MEAN EACH LEVEL FACTOR B (M3), AND SSB (S3)
0200 LET S3 = 0
0210 LET M3 = 0
0220 LET G = 0
0230 GOSUB 3010
0240 FOR K = 1 TO Q
0250 FOR J = 1 TO P
0260 FOR L = 1 TO R
0270 FOR I = 1 TO N
0280 READ S
0290 LET M3 = M3 + S/(N*P*R)
0300 NEXT I
0310 NEXT L
0320 LET H = N*R*(Q - 1)
0330 GOSUB 3050
0340 NEXT J
0350 LET S3 = S3 + (M3 - M)*(M3 - M)*N*P*R
0360 PRINT "MEAN B";K;" = ";M3
0370 GOSUB 3110
0380 LET M3 = 0
0390 LET G = G + N*R
0400 LET H = G
0410 GOSUB 3010
0420 GOSUB 3050
0430 NEXT K
0440 REM CALCULATE SSTOTAL (S1), MEAN EACH LEVEL
0450 REM FACTOR A (M2), SSA (S2), MEAN EACH LEVEL
0460 REM AB INTERACTION (M5), SSAB (S5), MEAN EACH LEVEL ABC
0470 REM INTERACTION (M8), VAR. EACH LEVEL ABC (V9), AND
0480 REM SSWITHIN CELLS (S9)
0490 LET S1 = 0
0500 LET S2 = 0
0510 LET S5 = 0
0520 LET S9 = 0
0530 LET M2 = 0
0540 LET M5 = 0
0550 LET M8 = 0
0560 LET V9 = 0
0570 LET G1 = 0
0580 LET G2 = 0
0590 FOR J = 1 TO P
0600 FOR K = 1 TO Q
0610 FOR L = 1 TO R
0620 GOSUB 3010
0630 IF G2 = 0 THEN 660
0640 LET H = G2
0650 GOSUB 3050
0660 IF G1 <> 0 THEN 750
0670 FOR I = 1 TO N
0680 READ S
0690 LET M2 = M2 + S/(N*Q*R)
0700 LET M5 = M5 + S/(N*R)
0710 LET M8 = M8 + S/N
0720 NEXT I
0730 LET G1 = 1
0740 GOTO 620
0750 FOR I = 1 TO N
```

```
0760 READ S
0770 LET S1 = S1 + (S - M)*(S - M)
0780 LET S9 = S9 + (S - M8)*(S - M8)
0790 LET V9 = V9 + (S - M8)*(S - M8)/(N - 1)
0800 NEXT I
0810 PRINT "MEAN CELL A";J;"B";K;"C";L;" = ";M8;
0820 PRINT "VAR. CELL A";J;"B";K;"C";L;" = ";V9
0830 LET V9 = 0
0840 GOSUB 3110
0850 LET M8 = 0
0860 LET G1 = 0
0870 LET G2 = G2 + N
0880 NEXT L
0890 LET S5 = S5 + (M5 - M)*(M5 - 5)*N*R - S3/(P*Q)
0900 PRINT "MEAN A";J;"B";K;" = ";M5
0910 GOSUB 3110
0920 LET M5 = 0
0930 NEXT K
0940 LET S2 = S2 + (M2 - M)*(M2 - M)*N*Q*R
0950 PRINT "MEAN A";J;" = ";M2
0960 GOSUB 3110
0970 LET M2 = 0
0980 NEXT J
0990 LET S5 = S5 - S2
1000 REM CALCULATE MEAN EACH LEVEL FACTOR C (M4), AND SSC (S4)
1010 LET G = 0
1020 LET S4 = 0
1030 LET M4 = 0
1040 GOSUB 3010
1050 FOR L = 1 TO R
1060 FOR J = 1 TO P
1070 FOR K = 1 TO Q
1080 FOR I = 1 TO N
1090 READ S
1100 LET M4 = M4 + S/(N*P*Q)
1110 NEXT I
1120 LET H = N*(R - 1)
1130 GOSUB 3050
1140 NEXT K
1150 NEXT J
1160 LET S4 = S4 + (M4 - M)*(M4 - M)*N*P*Q
1170 PRINT "MEAN C";L;" = ";M4
1180 GOSUB 3110
1190 LET M4 = 0
1200 GOSUB 3010
1210 LET G = G + N
1220 LET H = G
1230 GOSUB 3050
1240 NEXT L
1250 REM CALCULATE MEAN EACH LEVEL AC (M6), AND SSAC (S6)
1260 LET M6 = 0
1270 LET S6 = 0
1280 LET G = 0
1290 LET G1 = 0
1300 GOSUB 3010
1310 FOR J = 1 TO P
1320 FOR L = 1 TO R
1330 FOR K = 1 TO Q
1340 FOR I = 1 TO N
1350 READ S
1360 LET M6 = M6 + S/(N*Q)
1370 NEXT I
1380 LET H = N*R - N
1390 GOSUB 3050
1400 NEXT K
1410 LET S6 = S6 + (M6 - M)*(M6 - M)*N*Q - S2/(P*R) - S4/(P*R)
1420 PRINT "MEAN A";J;"C";L;" = ";M6
1430 GOSUB 3110
1440 LET M6 = 0
1450 LET G = G + N
1460 GOSUB 3010
```

Program 15. (cont.)

```
1470 LET H = G
1480 GOSUB 3050
1490 IF G1 = 0 THEN 1520
1500 LET H = G1
1510 GOSUB 3050
1520 NEXT L
1530 LET G = 0
1540 LET G1 = G1 + N*Q*R
1550 GOSUB 3010
1560 LET H = G1
1570 GOSUB 3050
1580 NEXT J
1590 REM CALC. MEAN EACH LEVEL BC (M7), SSBC (S7) AND SSABC (S8)
1600 LET S7 = 0
1610 LET S8 = 0
1620 LET M7 = 0
1630 LET M8 = 0
1640 LET G = 0
1650 GOSUB 3010
1660 FOR K = 1 TO Q
1670 FOR L = 1 TO R
1680 FOR J = 1 TO P
1690 FOR I = 1 TO N
1700 READ S
1710 LET M7 = M7 + S/(N*P)
1720 LET M8 = M8 + S/N
1730 NEXT I
1740 LET G1 = P*Q*R
1750 LET S8 = S8 + (M8 - M)*(M8 - M)*N
1760 LET S8 = S8 - S2/G1 - S3/G1 - S4/G1 - S5/G1 - S6/G1
1770 LET M8 = 0
1780 LET H = N*R*Q - N
1790 GOSUB 3050
1800 NEXT J
1810 PRINT "MEAN B";K;"C";L;" = ";M7
1820 GOSUB 3110
1830 LET S7 = S7 + (M7 - M)*(M7 - M)*N*P - S3/(Q*R) - S4/(Q*R)
1840 LET M7 = 0
1850 LET G = G + N
1860 GOSUB 3010
1870 LET H = G
1880 GOSUB 3050
1890 NEXT L
1900 NEXT K
2000 LET S8 = S8 - S7
2010 LET D1 = N*P*Q*R - 1
2020 LET D2 = P - 1
2030 LET D3 = Q - 1
2040 LET D4 = R - 1
2050 LET D5 = D2*D3
2060 LET D6 = D2*D4
2070 LET D7 = D3*D4
2080 LET D8 = D5*D4
2090 LET D9 = P*Q*R*(N - 1)
2110 REM CALCULATE MEAN EACH BLOCK (M0), SSBLOCKS (U1),
2120 REM SSAXBLOCKS (U2), AND SSABXBLOCKS (U5)
2130 REM FOR SPF-J.KL, SSCXS.W.GP (T2) FOR SPF-JK.L,
2140 REM SSBXS.W.GP (T4) FOR SPF-J.KL, SSAXBLOCKS
2150 REM (U2), AND SSABXBLOCKS (U5)
2160 LET M0 = 0
2170 LET M1 = 0
2180 LET M3 = 0
2190 LET T1 = 0
2200 LET T3 = 0
2210 LET U1 = 0
2220 LET G = 0
2230 GOSUB 3010
2240 FOR I = 1 TO N
2250 FOR J = 1 TO P
2260 FOR K = 1 TO Q
2270 FOR L = 1 TO R
```

Program 15. (cont.)

```
2280 READ S
2290 LET M0 = M0 + S/(P*Q*R)
2300 LET M1 = M1 + S/R
2310 LET M3 = M3 + S/(Q*R)
2320 LET H = N - 1
2330 GOSUB 3050
2340 NEXT L
2350 LET G1 = N*P*Q
2360 LET T1 = T1 + (M1 - M)*(M1 - M)*R - S2/G1 - S3/G1 - S5/G1
2380 LET M1 = 0
2390 NEXT K
2400 LET T3 = T3 + (M3 - M)*(M3 - M)*Q*R - S2/(N*P)
2410 LET M3 = 0
2420 NEXT J
2430 LET U1 = U1 + (M0 - M)*(M0 - M)*P*Q*R
2440 IF A < 3 THEN 2470
2450 PRINT "MEAN N";I;" = ";M0
2460 GOSUB 3110
2470 LET M0 = 0
2480 GOSUB 3010
2490 LET G = G + 1
2500 LET H = G
2510 GOSUB 3050
2520 NEXT I
2550 LET U2 = T3 - U1
2560 LET U5 = T1 - U1
2600 REM CALCULATE SSCXBLOCKS (U4)
2620 LET G = 0
2630 LET G1 = 0
2640 LET T5 = 0
2650 LET U4 = 0
2660 LET M4 = 0
2670 LET M5 = 0
2680 GOSUB 3010
2690 FOR L = 1 TO R
2700 FOR I = 1 TO N
2710 FOR J = 1 TO P
2720 FOR K = 1 TO Q
2730 READ S
2740 LET M4 = M4 + S/(P*Q)
2750 LET M5 = M5 + S/Q
2760 LET H = N*R - 1
2770 GOSUB 3050
2780 NEXT K
2790 LET G2 = N*P*R
2800 LET T5 = T5 + (M5 - M)*(M5 - M)*Q - S2/G2 - S4/G2 - S6/G2 - T3/G2
2810 LET M5 = 0
2820 NEXT J
2830 LET U4 = U4 + (M4 - M)*(M4 - M)*P*Q - S4/(N*R) - U1/(N*R)
2840 LET M4 = 0
2850 GOSUB 3010
2860 LET G = G + 1
2870 LET H = G
2880 GOSUB 3050
2890 NEXT I
2900 GOSUB 3010
2910 LET H = G
2920 GOSUB 3050
2930 NEXT L
2990 GOTO 3200
3000 REM RESTORE DATA POINTER AND ADVANCE DATA POINTER
3010 RESTORE
3020 READ S, S, S, S, S
3030 RETURN
3040 REM ADVANCE DATA POINTER
3050 FOR L1 = 1 TO H
3060 READ S
3070 IF S > 2E22 THEN 3090
3080 NEXT L1
3090 RETURN
3100 REM SUBROUTINE TO STOP EXECUTION UNTIL DATA ARE TRANSCRIBED
```

Program 15. (cont.)

```
3110 LET T = T + 1
3120 IF T <= 10 THEN RETURN
3130 PRINT "ENTER 1 FOLLOWED BY CARRIAGE RETURN TO CONTINUE."
3140 INPUT R1
3150 LET T = 0
3160 RETURN
3180 REM CALCULATE SSBXBLOCKS (U3), SSBCXBLOCKS (U7),
3190 REM SSABCXBLOCKS (U8), SSABXBLOCKS (U5), AND SSACXBLOCKS (U6)
3200 LET G = 0
3210 LET G1 = 0
3220 LET G2 = 0
3230 LET M3 = 0
3240 LET M7 = 0
3250 LET U3 = 0
3260 LET U7 = 0
3270 GOSUB 3010
3280 FOR K = 1 TO Q
3290 FOR I = 1 TO N
3300 FOR L = 1 TO R
3310 FOR J = 1 TO P
3320 READ S
3330 LET M3 = M3 + S/(P*R)
3340 LET M7 = M7 + S/P
3350 LET H = N*Q*R - 1
3360 GOSUB 3050
3370 NEXT J
3380 LET G3 = (M7 - M)*(M7 - M)*P - S3/(N*Q*R) - S4/(N*Q*R) - S7/(N*Q*R)
3390 LET U7 = U7 + G3 - U1/(N*Q*R) - U4/(N*Q*R)
3400 LET M7 = 0
3410 GOSUB 3010
3420 LET G = G + N
3430 LET H = G
3440 GOSUB 3050
3450 NEXT L
3460 LET U3 = U3 + (M3 - M)*(M3 - M)*P*R - S3/(N*Q) - U1/(N*Q)
3470 LET M3 = 0
3480 GOSUB 3010
3490 LET G1 = G1 + 1
3500 LET G = G1 + G2
3510 LET H = G
3520 GOSUB 3050
3530 NEXT I
3540 GOSUB 3010
3550 LET G1 = 0
3560 LET G2 = G2 + N*R
3570 LET G = G2
3580 LET H = G
3590 GOSUB 3050
3600 NEXT K
3610 LET U7 = U7 - U3
3620 LET U6 = T5 + T3 - U2 - U4 - U1
3630 LET U5 = T1 - U2 - U3 - U1
3640 LET U8 = S9 - U2 - U3 - U4 - U5 - U6 - U7 - U1
3650 LET F1 = N - 1
3660 LET F2 = (P - 1)*(N - 1)
3670 LET F3 = (Q - 1)*(N - 1)
3680 LET F4 = (R - 1)*(N - 1)
3690 LET F5 = F2*(Q - 1)
3700 LET F6 = F2*(R - 1)
3710 LET F7 = F4*(Q - 1)
3720 LET F8 = F7*(P - 1)
3740 REM F-TABLE FOR RBF-.JKL
3750 LET T = 13
3760 GOSUB 3120
3770 PRINT
3780 PRINT TAB(9);"SS";TAB(22);"DF";TAB(35);"MS";TAB(48);"F"
3790 PRINT "TOTAL";TAB(9);S1;TAB(22);D1
3800 PRINT "  BLOCK";TAB(9);U1;TAB(22);F1;TAB(35);U1/F1;TAB(48);
3810 PRINT (U1/F1)/((U2+U3+U4+U5+U6+U7+U8)/(F2+F3+F4+F5+F6+F7+F8))
3820 PRINT "  A";TAB(9);S2;TAB(22);D2;TAB(35);S2/D2;TAB(48);
```

```
3830 PRINT (S2/D2)/(U2/F2)
3840 PRINT "  AXBLK";TAB(9);U2;TAB(22);F2;TAB(35);U2/F2
3850 PRINT "  B";TAB(9);S3;TAB(22);D3;TAB(35);S3/D3;TAB(48);
3860 PRINT (S3/D3)/(U3/F3)
3870 PRINT "  BXBLK";TAB(9);U3;TAB(22);F3;TAB(35);U3/F3
3880 PRINT "  C";TAB(9);S4;TAB(22);D4;TAB(35);S4/D4;TAB(48);
3890 PRINT (S4/D4)/(U4/F4)
3900 PRINT "  CXBLK";TAB(9);U4;TAB(22);F4;TAB(35);U4/F4
3910 PRINT "  AB";TAB(9);S5;TAB(22);D5;TAB(35);S5/D5;TAB(48);
3920 PRINT (S5/D5)/(U5/F5)
3930 PRINT "  ABXBK";TAB(9);U5;TAB(22);F5;TAB(35);U5/F5
3940 PRINT "  AC";TAB(9);S6;TAB(22);D6;TAB(35);S6/D6;TAB(48);
3950 PRINT (S6/D6)/(U6/F6)
3960 PRINT "  ACXBK";TAB(9);U6;TAB(22);F6;TAB(35);U6/F6
3970 PRINT "  BC";TAB(9);S7;TAB(22);D7;TAB(35);S7/D7;TAB(48);
3980 PRINT (S7/D7)/(U7/F7)
3990 PRINT "  BCXBK";TAB(9);U7;TAB(22);F7;TAB(35);U7/F7
4000 PRINT "  ABC";TAB(9);S8;TAB(22);D8;TAB(35);S8/D8;TAB(48);
4010 PRINT (S8/D8)/(U8/F8)
4020 PRINT "  ABCXB";TAB(9);U8;TAB(22);F8;TAB(35);U8/F8
4030 PRINT "TOTAL";TAB(9);S2+S3+S4+S5+S6+S7+S8+S9;TAB(22);D1
4040 GOTO 9999
9998 DATA 2E23
9999 END
```

Program 16. Latin square design (requires equal n in all cells), LS-K.

```
0010 REM THIS PROGRAM IS FOR A LATIN SQUARE DESIGN
0020 REM (LS-K)
0030 REM SEE APPROPRIATE EXAMPLE FOR ENTERING DATA
0040 PRINT
0050 READ N, P
0060 REM CALCULATE GRAND MEAN
0070 GOSUB 3010
0080 LET M = 0
0090 FOR K = 1 TO P
0100 FOR J = 1 TO P
0110 READ R
0120 FOR I = 1 TO N
0130 READ R
0140 LET M = M + R/(N*P*P)
0150 NEXT I
0160 NEXT J
0170 NEXT K
0180 REM CALCULATE SSTOTAL (S1), MEANS FOR EACH CELL (M1)
0190 REM VAR. FOR EACH CELL (V1), SSWITHIN CELLS (S5),
0200 REM MEANS FOR LEVELS OF ROWS (M2), SSROWS (S2)
0210 LET M1 = 0
0220 LET M2 = 0
0230 LET S1 = 0
0240 LET S2 = 0
0250 LET S5 = 0
0260 LET V1 = 0
0270 LET T = 0
0280 LET L = 0
0290 GOSUB 3010
0300 FOR J = 1 TO P
0310 FOR K = 1 TO P
0320 READ R
0330 FOR I = 1 TO N
0340 READ R
0350 LET S1 = S1 + (R - M)*(R - M)
0360 LET M2 = M2 + R/(N*P)
0370 LET M1 = M1 + R/N
0380 NEXT I
0390 GOSUB 3010
0400 IF L <> 0 THEN GOSUB 3050
0410 READ R
0420 FOR I = 1 TO N
0430 READ R
0440 LET S5 = S5 + (R - M1)*(R - M1)
0450 LET V1 = V1 + (R - M1)*(R - M1)/(N - 1)
0460 NEXT I
0470 PRINT "MEAN CELL A";J;"B";K;" = ";M1;
0480 PRINT "VAR. CELL A";J;"B";K;" = ";V1
0490 GOSUB 3110
0500 LET M1 = 0
0510 LET V1 = 0
0520 LET L = L + N + 1
0530 GOSUB 3010
0540 GOSUB 3050
0550 NEXT K
0560 LET S2 = S2 + (M2 - M)*(M2 - M)*N*P
0570 PRINT "MEAN LEVEL A (ROW)";J;" = ";M2
0580 GOSUB 3110
0590 LET M2 = 0
0600 NEXT J
0610 REM CALCULATE MEANS FOR LEVELS OF COLUMNS (M3),
0620 REM AND SSCOLUMNS (S3)
0630 LET L = 0
0640 LET M3 = 0
0650 LET S3 = 0
0660 GOSUB 3010
0670 FOR K = 1 TO P
0680 FOR J = 1 TO P
0690 READ R
0700 FOR I = 1 TO N
0710 READ R
```

Program 16. (cont.)

```
0720 LET M3 = M3 + R/(N*P)
0730 NEXT I
0740 LET L = (N + 1)*(P - 1)
0750 GOSUB 3050
0760 NEXT J
0770 PRINT "MEAN COLUMN ";K;" = ";M3
0780 LET S3 = S3 + (M3 - M)*(M3 - M)*N*P
0790 LET M3 = 0
0800 LET L2 = L2 + N + 1
0810 GOSUB 3010
0820 LET L = L2
0830 GOSUB 3050
0840 NEXT K
0850 REM CALCULATE MEANS FOR LEVELS OF TREATMENTS-C (M4),
0860 REM AND SSTREATMENTS-C (S4)
0870 LET M4 = 0
0880 LET S4 = 0
0890 FOR L2 = 1 TO P
0900 GOSUB 3010
0910 FOR J = 1 TO P*P
0920 READ R
0930 IF R = L2 THEN 970
0940 LET L = N
0950 GOSUB 3050
0960 GOTO 1010
0970 FOR I = 1 TO N
0980 READ R
0990 LET M4 = M4 + R/(N*P)
1000 NEXT I
1010 NEXT J
1020 PRINT "MEAN TREATMENT-C";L2;" = ";M4
1030 GOSUB 3110
1040 LET S4 = S4 + (M4 - M)*(M4 - M)*N*P
1050 LET M4 = 0
1060 NEXT L2
1070 LET S6 = S1 - S2 - S3 - S4 - S5
1080 LET D1 = N*P*P - 1
1090 LET D2 = P - 1
1100 LET D3 = D2
1110 LET D4 = D2
1120 LET D5 = P*P*(N - 1)
1130 LET D6 = (P - 1)*(P - 2)
1140 LET T = 13
1150 GOSUB 3120
1160 PRINT
1170 PRINT TAB(9);"SS";TAB(22);"DF";TAB(35);"MS";TAB(48);"F"
1180 PRINT "TOTAL";TAB(9);S1;TAB(22);D1
1190 PRINT "BET A";TAB(9);S2;TAB(22);D2;TAB(35);S2/D2;
1200 PRINT TAB(48);(S2/D2)/(S5/D5)
1210 PRINT "BET B";TAB(9);S3;TAB(22);D3;TAB(35);S3/D3;
1220 PRINT TAB(48);(S3/D3)/(S5/D5)
1230 PRINT "BET C";TAB(9);S4;TAB(22);D4;TAB(35);S4/D4;
1240 PRINT TAB(48);(S4/D4)/(S5/D5)
1250 PRINT "RES.";TAB(9);S6;TAB(22);D6;TAB(35);S6/D6;
1260 PRINT TAB(48);(S6/D6)/(S5/D5)
1270 PRINT "W CEL";TAB(9);S5;TAB(22);D5;TAB(35);S5/D5
1280 GOTO 9999
3000 REM RESTORE DATA POINTER AND ADVANCE DATA POINTER
3010 RESTORE
3020 READ R, R
3030 RETURN
3040 REM ADVANCE DATA POINTER
3050 FOR L1 = 1 TO L
3060 READ R
3070 IF R > 2E22 THEN 3090
3080 NEXT L1
3090 RETURN
3100 REM SUBROUTINE TO STOP EXECUTION UNTIL DATA ARE TRANSCRIBED
3110 LET T = T + 1
3120 IF T <= 12 THEN RETURN
3130 PRINT "ENTER 1 FOLLOWED BY CARRIAGE RETURN TO CONTINUE."
```

Program 16. (cont.)

```
3140 INPUT R1
3150 LET T = 0
3160 RETURN
9998 DATA 2E23
9999 END
```

Program 17. Analysis of covariance--one factor (requires equal n in all cells).

```
0010 REM THIS PROGRAM IS FOR AN ANALYSIS OF
0020 REM COVARIANCE FOR A RANDOMIZED GROUPS
0030 REM DESIGN
0040 REM SEE APPROPRIATE EXAMPLE FOR ENTERING DATA
0050 PRINT
0060 READ N,T
0070 REM CALCULATE GRAND MEANS
0080 LET X = 0
0090 LET Y = 0
0100 FOR I = 1 TO N*T
0110 READ Q, Q1
0120 LET X = X + Q/(N*T)
0130 LET Y = Y + Q1/(N*T)
0140 NEXT I
0150 REM CALCULATE MEANS FOR X TREATMENTS (X1), MEANS FOR
0160 REM Y TREATMENTS (Y1), SSTOTX (S1), SSTOTY(T1),
0170 REM SSTOTXY (W1), SSBET. TREAT X (S2), SSBET. TREAT
0180 REM Y (T2), SSBET. TREAT XY (W2), SSWITHIN CELL
0190 REM FOR X GROUPS (S3), SSWITHIN CELL FOR Y GROUPS
0200 REM (T3), SSWITHIN CELL FOR XY GROUPS (W3),
0210 REM WITHIN CELL X (S4), SSWITHIN CELL Y (T4),
0220 REM SSWITHIN CELL XY (W4),AND SUM (W3*W3/S3)'
0230 LET U1 = 0
0240 LET X1 = 0
0250 LET Y1 = 0
0260 LET S1 = 0
0270 LET S2 = 0
0280 LET S3 = 0
0290 LET S4 = 0
0300 LET T1 = 0
0310 LET T2 = 0
0320 LET T3 = 0
0330 LET W1 = 0
0340 LET W2 = 0
0350 LET W3 = 0
0360 LET W4 = 0
0370 LET A1 = 0
0380 LET L = 0
0390 GOSUB 3010
0400 FOR J = 1 TO T
0410 FOR I = 1 TO N
0420 READ Q, Q1
0430 LET X1 = X1 + Q/N
0440 LET Y1 = Y1 + Q1/N
0450 NEXT I
0460 PRINT "MEAN TREAT. X";J;" = ";X1;
0470 PRINT "MEAN TREAT. Y";J;" = ";Y1'
0480 GOSUB 3110
0490 GOSUB 3010
0500 IF L = 0 THEN 530
0510 LET H = L
0520 GOSUB 3050
0530 FOR I = 1 TO N
0540 READ Q1, Q
0550 LET S1 = S1 + (Q1 - X)*(Q1 - X)
0560 LET T1 = T1 + (Q - Y)*(Q - Y)
0570 LET W1 = W1 + (Q1 - X)*(Q - Y)
0580 LET S3 = S3 + (Q1 - X1)*(Q1 - X1)
0590 LET T3 = T3 + (Q - Y1)*(Q - Y1)
0600 LET W3 = W3 + (Q1 - X1)*(Q - Y1)
0610 NEXT I
0620 LET S4 = S4 + S3
0630 LET T4 = T4 + T3
0640 LET W4 = W4 + W3
0650 PRINT "SSWITH X ";J;" = ";S3
0660 PRINT "SSWITH Y ";J;" = ";T3
0670 PRINT "SSWITH XY ";J;" = ";W3
0680 GOSUB 3110
0690 LET A1 = A1 + W3*W3/S3
0700 LET S3 = 0
```

Program 17. (cont.)

```
0710 LET T3 = 0
0720 LET W3 = 0
0730 LET S2 = S2 + (X1 - X)*(X1 - X)*N
0740 LET T2 = T2 + (Y1 - Y)*(Y1 - Y)*N
0750 LET W2 = W2 + (X1 - X)*(Y1 - Y)*N
0760 LET X1 = 0
0770 LET Y1 = 0
0780 LET L = L + N
0790 LET H = L
0800 GOSUB 3010
0810 GOSUB 3050
0820 NEXT J
0830 LET U1 = 13
0840 GOSUB 3110
0850 PRINT
0860 PRINT TAB(9);"SUMX*X";TAB(22);"SUMY*Y";TAB(35);"SUMX*Y"
0870 PRINT "TOTAL";TAB(9);S1;TAB(22);T1;TAB(35);W1
0880 PRINT "BET.";TAB(9);S2;TAB(22);T2;TAB(35);W2
0890 PRINT "WITH.";TAB(9);S4;TAB(22);T4;TAB(35);W4
0900 PRINT
0910 REM CALCULATE A1 = T1 - (SUM(SUMX*Y/SUMX*X))
0920 LET A1 = T4 - A1
0930 REM CALCULATE A2 = T4 - W4*W4/S4
0940 LET A2 = T4 - W4*W4/S4
0950 REM F-TEST OF DIFFERENCES BETWEEN
0960 REM GROUP REGRESSION COEFFICIENTS
0970 PRINT "F-TEST OF DIFFERENCES BETWEEN"
0980 PRINT "GROUP REGRESSION COEFFICIENTS"
0990 PRINT "F = ";((A2 - A1)/(T - 1))/(A1/(T*(N - 2)))
1000 PRINT "WITH ";T - 1;" AND ";T*(N - 2); " DEGREES OF FREEDOM"
1010 PRINT "IF ABOVE F-RATIO IS SIGNIFICANT DO NOT CONTINUE"
1020 PRINT "WITH ANALYSIS OF COVARIANCE."
1030 LET U1 = 13
1040 GOSUB 3110
1050 REM CALCULATE A4 = T1 - W1*W1/S1 - A2
1060 LET A4 = T1 - W1*W1/S1 - A2
1070 PRINT
1080 PRINT "F-RATIO FOR ANALYSIS OF COVARIANCE"
1090 PRINT "F = ";(A4/(T - 1))/(A2/(T*(N - 1) - 1))
1100 PRINT "WITH ";T - 1; " AND ";T*(N - 1) - 1;" DEGREES OF FREEDOM"
1110 GOTO 9999
3000 REM RESTORE DATA POINTER AND ADVANCE TO FIRST DATA POINT
3010 RESTORE
3020 READ Q,Q
3030 RETURN
3040 REM SUBROUTINE TO ADVANCE POINTER TO NEXT OBSERVATION
3050 FOR L1 = 1 TO H
3060 READ Q,Q
3070 IF R > 2E22 THEN 3090
3080 NEXT L1
3090 RETURN
3100 REM SUBROUTINE TO STOP EXECUTION UNTIL DATA ARE TRANSCRIBED
3110 LET U1 = U1 + 1
3120 IF U1 < 10 THEN RETURN
3130 PRINT
3140 PRINT "ENTER 1 FOLLOWED BY A CARRIAGE RETURN TO CONTINUE."
3150 INPUT Q3
3160 LET U1 = 0
3170 RETURN
9998 DATA 2E23
9999 END
```

Chapter 5

Difference Formulas Used in Analysis of Variance Programs

Difference formulas for sums of squares used in the analysis of variance programs are presented in this chapter. Formulas for degrees of freedom are presented next to the difference formulas. A review of the summation notation used in the formulas can be found in Hays (1981, pp. 616-624). Unless noted under the formulas for a given analysis the following definitions apply to the terms in the analyses:

The first factor in an analysis is designated as j.

The second factor (if there is a second factor) is designated as k.

The third factor (if there is a third factor) is designated as l.

Blocks (described as subjects in some studies) are designated as i.

An individual observation in a one factor analysis is x_{ij}.

An individual observation in a two factor analysis is x_{ijk}.

An individual observation in a three factor analysis is x_{ijkl}.

The mean of all the observations in an analysis (grand mean) is designated as $\bar{X}$.

A mean for a level of the first factor is $\bar{X}_j$.

A mean for a level of the second factor (if there is a second factor) is $\bar{X}_k$.

A mean for a level of the third factor (if there is a third

factor) is $\bar{X}_{l}$.

A mean for a level of the interaction involving the first and second factor is $\bar{X}_{jk}$.

A mean for a level of the interaction involving the first and third factor is $\bar{X}_{jl}$.

A mean for a level of the interaction involving the second and third factor is $\bar{X}_{kl}$.

A mean for a level of the triple interaction (first factor by second factor by third factor) is $\bar{X}_{jkl}$.

If a block is represented in all factors, the block mean is $\bar{X}_{i}$.

A mean for a block at a level of the first factor is $\bar{X}_{ij}$.

A mean for a block at a level of the second factor is $\bar{X}_{ik}$.

A mean for a block at a level of the third factor is $\bar{X}_{il}$.

A mean for a block at a level of the interaction of the first and second factor is $\bar{X}_{ijk}$.

A mean for a block at a level of the interaction of the first and third factor is $\bar{X}_{ijl}$.

A mean for a block at a level of the interaction of the second and third factor is $\bar{X}_{ikl}$.

The analysis of covariance requires two sets of observations, x observations and y observations. Formulas involving y observations have y substituted for x and $\bar{Y}$ substituted for $\bar{X}$.

Formulas for Program 1. Completely randomized analysis of variance (does not require equal n in all cells), CR-K

	SS	df
Total	$\sum^{J}\sum^{I}(x_{ij} - \bar{X})^2$	$N - 1$
Between groups	$\sum^{J} i_j(\bar{X}_j - \bar{X})^2$	$J - 1$
Within groups	$\sum^{J}\sum^{I}(x_{ij} - \bar{X}_j)^2$	$(i - 1)j$

The formulas shown above do not require an equal number of observations for each group.

Formulas for Program 2. Completely randomized factorial analysis of variance (does not require equal n in all cells), CRF-JK.

	SS	df
A	$k\sum^{J}(\bar{X}_j - \bar{X})^2$	$j - 1$
B	$j\sum^{K}(\bar{X}_k - \bar{X})^2$	$k - 1$
AB	$\sum\sum^{KJ}(\bar{X}_{jk} - \bar{X})^2$	$(j - 1)(k - 1)$
	$-SSA - SSB$	

The mean square within cell term is calculated from the raw data with the formula:

$$\frac{1}{jk}\sum\sum^{KJ}\frac{1}{i_{jk}}\;\frac{\sum\sum\sum^{KJI}(x_{ijk} - \bar{X}_{jk})^2}{\sum\sum^{KJ} i_{jk} - jk}$$

where $\frac{1}{jk}$ is the reciprocal of the number of cells in the analysis of variance (i.e., the reciprocal of the product of the number of levels of the first factor times the number of levels of the second factor). $\sum\sum^{KJ}\frac{1}{i_{jk}}$ is obtained by taking the reciprocal of the number of observations in each cell (separately for each cell) and summing these reciprocals. $\sum\sum^{KJ} i_{jk}$ is the number of observations in each cell summed across all cells. x_{ijk} is an individual observation. $\bar{X}_{jk}$ is the mean for one of the cells in the analysis of variance (i.e., a cell at a given level of the first factor and a given level of the second factor. A table is made of all the $\bar{X}_{jk}$ means. Then $\bar{X}_j$, $\bar{X}_k$ and $\bar{X}$ are calculated from the means in the table, not from the individual observations. The formulas shown above do not require an equal number of observations for each cell.

Formulas for Program 3. Completely randomized factorial analysis of variance (does not require equal n in all cells), CRF-JKL.

	SS	df
A	$kl\sum^{J}(\bar{X}_j - \bar{X})^2$	$j - 1$
B	$jl\sum^{K}(\bar{X}_k - \bar{X})^2$	$k - 1$
C	$jk\sum^{L}(\bar{X}_l - \bar{X})^2$	$l - 1$
AB	$l\sum\sum^{KJ}(\bar{X}_{jk} - \bar{X})^2$ -SSA -SSB	$(j - 1)(k - 1)$
AC	$k\sum\sum^{LJ}(\bar{X}_{jl} - \bar{X})^2$ -SSA -SSC	$(j - 1)(l - 1)$
BC	$j\sum\sum^{LK}(\bar{X}_{kl} - \bar{X})^2$ -SSB -SSC	$(k - 1)(l - 1)$
ABC	$\sum\sum\sum^{LKJ}(\bar{X}_{jkl} - \bar{X})^2$ -SSA -SSB -SSC -SSAB - SSAC -SSBC	$(j - 1)(k - 1)(l - 1)$

The mean square within cell term is calculated from raw data with the formula:

$$\frac{1}{jkl}\sum\sum\sum^{LKJ}\frac{1}{i_{jkl}}\;\frac{\sum\sum\sum\sum^{LKJI}(x_{ijkl} - \bar{X}_{jkl})^2}{\sum\sum\sum^{LKJ} i_{jkl} - jkl}$$

where $\frac{1}{jkl}$ is the reciprocal of the number of cells in the analysis of variance (i.e., the reciprocal of the product of the number of levels of the first factor times the number of levels of the second factor times the number of levels of the third factor). $\sum\sum\sum^{LKJ}\frac{1}{i_{jkl}}$ is obtained by taking the reciprocal of the number of observations in each cell (i.e., separately for each cell) and summing these reciprocals. $\sum\sum\sum^{LKJ} i_{jkl}$ is the number of observations in each cell summed across all cells. x_{ijkl} is an individual observation. $\bar{X}_{jkl}$ is the mean for one of the cells in the analysis of variance (i.e., a cell at a given level of the first factor, the second factor, and third factor). A table is made of all the $\bar{X}_{jkl}$ means. Then $\bar{X}_j$, $\bar{X}_k$, $\bar{X}_l$, $\bar{X}_{jk}$, $\bar{X}_{jl}$, $\bar{X}_{kl}$, and $\bar{X}$ are calculated from the means in the table, not from the individual observations. The formulas shown above do not require an equal number of observations for each cell.

Formulas for Programs 4 and 7. Completely randomized analysis of variance (requires equal n in all cells), CR-K.

	SS	df
Total	$\sum^{J}\sum^{I}(x_{ij} - \bar{X})^2$	$ij - 1$
Between groups	$i\sum^{J}(\bar{X}_j - \bar{X})^2$	$j - 1$
Within groups	$\sum^{J}\sum^{K}(x_{ij} - \bar{X}_j)^2$	$(i - 1)j$

The formulas shown above require an equal number of observations for each group.

Formulas for Programs 4 and 8. Randomized block analysis of variance (requires equal n in all cells), RB-K.

	SS	df
Total	$\sum^{J}\sum^{I}(x_{ij} - \bar{X})^2$	$ij - 1$
Between blocks	$j\sum^{I}(\bar{X}_i - \bar{X})^2$	$i - 1$
Between treatments	$i\sum^{J}(\bar{X}_j - \bar{X})^2$	$j - 1$
Residual	$\sum^{J}\sum^{I}(x_{ij} - \bar{X}_i - \bar{X}_j + \bar{X})^2$	$(i - 1)(j - 1)$
	or $\sum^{J}\sum^{I}(x_{ij} - \bar{X})^2$	
	-SSBetween blocks	
	-SSBetween treatments	

The formulas shown above require an equal number of observations for each cell.

Formulas for Programs 5 and 9. *Completely randomized factorial analysis of variance (requires equal n in all cells), CRF-JK.*

	SS	df
Total	$\sum^{K}\sum^{J}\sum^{I}(x_{ijk} - \bar{X})^2$	$ijk - 1$
A	$ik\sum^{J}(\bar{X}_j - \bar{X})^2$	$j - 1$
B	$ij\sum^{K}(\bar{X}_k - \bar{X})^2$	$k - 1$
AB	$i\sum^{K}\sum^{J}(\bar{X}_{jk} - \bar{X})^2$ -SSA -SSB	$(j - 1)(k - 1)$
Within cell	$\sum^{K}\sum^{J}\sum^{I}(x_{ijk} - \bar{X}_{jk})^2$	$(i - 1)jk$

The formulas shown above require an equal number of observations for each cell.

Formulas for Programs 5 and 10. Split-plot analysis of variance (requires equal n in all cells), SPF-J.K.

	SS	df
Total	$\sum^{K}\sum^{J}\sum^{I}(x_{ijk} - \bar{X})^2$	$ijk - 1$
Between	$k\sum^{J}\sum^{I}(\bar{X}_{ij} - \bar{X})^2$	$ij - 1$
A	$ik\sum^{J}(\bar{X}_{j} - \bar{X})^2$	$j - 1$
Sub. w. groups	$k\sum^{J}\sum^{I}(\bar{X}_{ij} - \bar{X}_{j})^2$	$(i - 1)j$
Within	$\sum^{K}\sum^{J}\sum^{I}(x_{ijk} - \bar{X}_{ij})^2$	$(k - 1)ij$
B	$ij\sum^{K}(\bar{X}_{k} - \bar{X})^2$	$k - 1$
AB	$i\sum^{K}\sum^{J}(\bar{X}_{jk} - \bar{X})^2$ −SSA −SSB	$(j - 1)(k - 1)$
B x sub. w. groups	$\sum^{K}\sum^{J}\sum^{I}(x_{ijk} - \bar{X}_{jk})^2$ − Sub. w. groups	$(i - 1)(k - 1)j$

The formulas shown above require an equal number of observations for each cell.

Formulas for Programs 5 and 11. Randomized block factorial analysis of variance (requires equal n in all cells), RBF-.JK.

	SS	df
Total	$\sum^{K}\sum^{J}\sum^{I}(x_{ijk} - \bar{X})^2$	ijk - 1
Blocks	$jk\sum^{I}(\bar{X}_i - \bar{X})^2$	i - 1
A	$ik\sum^{J}(\bar{X}_j - \bar{X})^2$	j - 1
A x blocks	$k\sum^{J}\sum^{I}(\bar{X}_{ij} - \bar{X}_j)^2$ -SSBlocks	(i - 1)(j - 1)
B	$ij\sum^{K}(\bar{X}_k - \bar{X})^2$	k - 1
B x blocks	$j\sum^{K}\sum^{I}(\bar{X}_{ik} - \bar{X}_k)^2$ -SSBlocks	(i - 1)(k - 1)
AB	$i\sum^{K}\sum^{J}(\bar{X}_{jk} - \bar{X})^2$ -SSA -SSB	(j - 1)(k - 1)
AB x blocks	$\sum^{K}\sum^{J}\sum^{I}(x_{ijk} - \bar{X}_{jk})^2$ -SSA x blocks -SSB x blocks -SSBlocks	(i - 1)(j - 1)(k - 1)

The formulas shown above require an equal number of observations for each cell.

Formulas for Programs 6 and 12. Completely randomized factorial analysis of variance (requires equal n in all cells), CRF-JKL.

	SS	df
Total	$\sum^{L}\sum^{K}\sum^{J}\sum^{I}(x_{ijkl} - \bar{X})^2$	$ijkl - 1$
A	$ikl\sum^{J}(\bar{X}_j - \bar{X})^2$	$j - 1$
B	$ijl\sum^{K}(\bar{X}_k - \bar{X})^2$	$k - 1$
C	$ijk\sum^{L}(\bar{X}_l - \bar{X})^2$	$l - 1$
AB	$il\sum^{K}\sum^{J}(\bar{X}_{jk} - \bar{X})^2$ $- SSA - SSB$	$(j - 1)(k - 1)$
AC	$ik\sum^{L}\sum^{J}(\bar{X}_{jl} - \bar{X})^2$ $- SSA - SSC$	$(j - 1)(l - 1)$
BC	$ij\sum^{L}\sum^{K}(\bar{X}_{kl} - \bar{X})^2$ $- SSB - SSC$	$(k - 1)(l - 1)$
ABC	$i\sum^{L}\sum^{K}\sum^{J}(\bar{X}_{jkl} - \bar{X})^2$ $- SSA - SSB - SSC - SSAB$ $- SSAC - SSBC$	$(j - 1)(k - 1)(l - 1)$
Within cell	$\sum^{L}\sum^{K}\sum^{J}\sum^{I}(x_{ijkl} - \bar{X}_{jkl})^2$	$(i - 1)jkl$

The formulas shown above require an equal number of observations for each cell.

Formulas for Programs 6 and 13. Split-plot analysis of variance (requires equal n in all cells), SPF-JK.L.

	SS	df
Total	$\sum^{L}\sum^{K}\sum^{J}\sum^{I}(x_{ijkl} - \bar{X})^2$	ijkl − 1
Between	$l\sum^{K}\sum^{J}\sum^{I}(\bar{X}_{ijk} - \bar{X})^2$	ijk − 1
A	$ikl\sum^{J}(\bar{X}_{j} - \bar{X})^2$	j − 1
B	$ijl\sum^{K}(\bar{X}_{k} - \bar{X})^2$	k − 1
AB	$il\sum^{K}\sum^{J}(\bar{X}_{jk} - \bar{X})^2$ −SSA − SSB	(j − 1)(k − 1)
Sub. w. groups	$l\sum^{K}\sum^{J}\sum^{I}(\bar{X}_{ijk} - \bar{X}_{jk})^2$	(i − 1)jk
Within	$\sum^{L}\sum^{K}\sum^{J}\sum^{I}(x_{ijkl} - \bar{X}_{ijk})^2$	(l − 1)ijk
C	$ijk\sum^{L}(\bar{X}_{l} - \bar{X})^2$	l − 1
AC	$ik\sum^{L}\sum^{J}(\bar{X}_{jl} - \bar{X})^2$ −SSA −SSC	(j − 1)(l − 1)
BC	$ij\sum^{L}\sum^{K}(\bar{X}_{kl} - \bar{X})^2$ −SSB −SSC	(k − 1)(l − 1)
ABC	$i\sum^{L}\sum^{K}\sum^{J}(\bar{X}_{jkl} - \bar{X})^2$ −SSA −SSB −SSC −SSAB −SSAC −SSBC	(j − 1)(k − 1)(l − 1)
C x sub. w. groups	$\sum^{L}\sum^{K}\sum^{J}\sum^{I}(x_{ijkl} - \bar{X}_{jkl})^2$ −SSSub. w. groups	(i − 1)(l − 1)jk

The formulas shown above require an equal number of observations for each cell.

Formulas for Programs 6 and 14. Split-plot analysis of variance (requires equal n in all cells), SPF-JK.L.

	SS	df
Total	$\sum^{L}\sum^{K}\sum^{J}\sum^{I}(x_{ijkl} - \bar{X})^2$	ijkl - 1
Between	$kl\sum^{J}\sum^{I}(\bar{X}_{ij} - \bar{X})^2$	ij - 1
A	$ikl\sum^{J}(\bar{X}_{j} - \bar{X})^2$	j - 1
Sub. w. groups	$kl\sum^{J}\sum^{I}(\bar{X}_{ij} - \bar{X}_{j})^2$	(i - 1)j
Within	$\sum^{L}\sum^{K}\sum^{J}\sum^{I}(x_{ijkl} - \bar{X}_{ij})^2$	(kl - 1)ij
B	$ijl\sum^{K}(\bar{X}_{k} - \bar{X})^2$	k - 1
AB	$il\sum^{K}\sum^{J}(\bar{X}_{jk} - \bar{X})^2$ -SSA -SSB	(j - 1)(k - 1)
B x sub. w. groups	$l\sum^{K}\sum^{J}\sum^{I}(\bar{X}_{ijk} - \bar{X}_{jk})^2$ -SSSub. w. groups	(i - 1)(k - 1)j
C	$ijk\sum^{L}(\bar{X}_{l} - \bar{X})^2$	l - 1
AC	$ik\sum^{L}\sum^{J}(\bar{X}_{jl} - \bar{X})^2$ -SSA -SSC	(j - 1)(l - 1)
C x sub. w. groups	$k\sum^{L}\sum^{J}\sum^{I}(\bar{X}_{ijl} - \bar{X}_{jl})^2$ -SSSub. w. groups	(i - 1)(l - 1)j
BC	$ij\sum^{L}\sum^{K}(\bar{X}_{kl} - \bar{X})^2$ -SSB -SSC	(k - 1)(l - 1)

ABC	$\sum^{L}\sum^{K}\sum^{J} i(\bar{X}_{jkl} - \bar{X})^2$	$(j - 1)(k - 1)(l - 1)$
	$-SSA - SSB - SSC - SSAB$	
	$-SSAC - SSBC$	
BC x sub. w. groups	$\sum^{L}\sum^{K}\sum^{J}\sum^{I}(x_{ijkl} - \bar{X}_{jkl})^2$	$(i - 1)(k - 1)(l - 1)j$
	$-$SSB x sub. w. groups	
	$-$SSC x sub. w. groups	
	$-$SSSub. w. groups	

The formulas shown above require an equal number of observations for each cell.

Formulas for Programs 6 and 15. Randomized block factorial analysis of variance (requires equal n in all cells), RBF-.JKL.

	SS	df
Total	$\sum^{L}\sum^{K}\sum^{J}\sum^{I}(x_{ijkl} - \bar{X})^2$	ijkl - 1
Blocks	$jkl\sum^{I}(\bar{X}_i - \bar{X})^2$	i - 1
A	$ikl\sum^{J}(\bar{X}_j - \bar{X})^2$	j - 1
A x blocks	$kl\sum^{J}\sum^{I}(\bar{X}_{ij} - \bar{X}_j)^2$ -SSBlocks	(i - 1)(j - 1)
B	$ijl\sum^{K}(\bar{X}_k - \bar{X})^2$	k - 1
B x blocks	$jl\sum^{K}\sum^{I}(\bar{X}_{ik} - \bar{X}_k)^2$ -SSBlocks	(i - 1)(k - 1)
C	$ijk\sum^{L}(\bar{X}_l - \bar{X})^2$	l - 1
C x blocks	$jk\sum^{L}\sum^{I}(\bar{X}_{il} - \bar{X}_l)^2$ -SSBlocks	(i - 1)(l - 1)
AB	$il\sum^{K}\sum^{J}(\bar{X}_{jk} - \bar{X})^2$ -SSA - SSB	(j - 1)(k - 1)
AB x blocks	$l\sum^{K}\sum^{J}\sum^{I}(\bar{X}_{ijk} - \bar{X}_{jk})^2$ -SSA x blocks -SSB x blocks -SSBlocks	(i - 1)(j - 1)(k - 1)
AC	$ik\sum^{L}\sum^{J}(\bar{X}_{jl} - \bar{X})^2$ -SSA - SSC	(j - 1)(l - 1)

Formulas for Programs 6 and 15 (cont.).

AC x blocks	$k\overset{LJI}{\Sigma\Sigma\Sigma}(\bar{X}_{ijl} - \bar{X}_{jl})^2$ -SSA x blocks -SSC x blocks -SSBlocks	(i - 1)(j - 1)(l - 1)
BC	$ij\overset{LK}{\Sigma\Sigma}(\bar{X}_{kl} - \bar{X})^2$ -SSB -SSC	(k - 1)(l - 1)
BC x blocks	$j\overset{LKI}{\Sigma\Sigma\Sigma}(\bar{X}_{ikl} - \bar{X}_{kl})^2$ -SSB x blocks -SSC x blocks -SSBlocks	(i - 1)(k - 1)(l - 1)
ABC	$i\overset{LKJ}{\Sigma\Sigma\Sigma}(\bar{X}_{jkl} - \bar{X})^2$ -SSA -SSB -SSC -SSAB -SSAC -SSBC	(j - 1)(k - 1)(l - 1)
ABC x blocks	$\overset{LKJI}{\Sigma\Sigma\Sigma\Sigma}(x_{ijkl} - \bar{X}_{jkl})^2$ -SSA x blocks -SSB x blocks -SSC x blocks -SSAB x blocks -SSAC x blocks -SSBC x blocks -SSBlocks	(i - 1)(j - 1)(k - 1)(l - 1)

The formulas shown above require an equal number of observations for each cell.

Formulas for Program 16. Latin square design (requires equal n in all cells), LS-K.

	SS	df
Total	$\sum^{L}\sum^{K}\sum^{J}\sum^{I}(x_{ijkl} - \bar{X})^2$	$ip^3 - 1$
Between A	$ip\sum^{J}(\bar{X}_j - \bar{X})^2$	$p - 1$
Between B	$ip\sum^{K}(\bar{X}_k - \bar{X})^2$	$p - 1$
Between C	$ip\sum^{L}(\bar{X}_l - \bar{X})^2$	$p - 1$
Residual	SSTotal - SSA -SSB -SSC -SSWithin cell	$(p - 1)(p - 2)$
Within cell	$\sum^{L}\sum^{K}\sum^{J}\sum^{I}(x_{ijkl} - \bar{X}_{jkl})^2$	$p^2(i - 1)$

The formulas shown above require an equal number of observations for each cell.

Formulas for Program 17. Analysis of covariance--one factor (requires equal n in all cells).

	SSX	SSY	SSXY	df
Total	$\sum^{J}\sum^{I}(x_{ij} - \bar{X})^2$	$\sum^{J}\sum^{I}(y_{ij} - \bar{Y})^2$	$\sum^{J}\sum^{I}(x_{ij} - \bar{X})(y_{ij} - \bar{Y})$	ij - 1
Within group 1	$\sum^{I}(x_{i1} - \bar{X}_1)^2$	$\sum^{I}(y_{i1} - \bar{Y}_1)^2$	$\sum^{I}(x_{i1} - \bar{X}_1)(y_{i1} - \bar{Y}_1)$	i - 1
Within group 2	$\sum^{I}(x_{i2} - \bar{X}_2)^2$	$\sum^{I}(y_{i2} - \bar{Y}_2)^2$	$\sum^{I}(x_{i2} - \bar{X}_2)(y_{i2} - \bar{Y}_2)$	i - 1
⋮	⋮	⋮	⋮	⋮
Within group n	$\sum^{I}(x_{ij} - \bar{X}_j)^2$	$\sum^{I}(y_{ij} - \bar{Y}_j)^2$	$\sum^{I}(x_{ij} - \bar{X}_j)(y_{ij} - \bar{Y}_j)$	i - 1
Within	$\sum^{J}\sum^{I}(x_{ij} - \bar{X}_j)^2$	$\sum^{J}\sum^{I}(y_{ij} - \bar{Y}_j)^2$	$\sum^{J}\sum^{I}(x_{ij} - \bar{X}_j)(y_{ij} - \bar{Y}_j)$	(i - 1)j
Between	$i\sum^{J}(\bar{X}_j - \bar{X})^2$	$i\sum^{J}(\bar{Y}_j - \bar{Y})^2$	$i\sum^{J}(\bar{X}_j - \bar{X})(\bar{Y}_j - \bar{Y})$	j - 1

$$F = \frac{\Sigma 4/(k - 1)}{\Sigma 1/[k(i - 2)]}$$

If F is significant with k - 1 and k(i - 1) degrees of freedom, the regression lines are not parallel and the results of the next F-ratio may not be valid.

$$F = \frac{\Sigma 5/(j - 1)}{\Sigma 2/[j(i - 1) - 1]}$$

If F is significant with j - 1 and j(i - 1) degrees of freedom, Y means differ after adjustment for X values.

The formulas shown above require an equal number of observations for each cell.

Formulas for Program 17 (cont.).

$$\Sigma 1 = \text{SSWithin } Y - \left[\frac{\text{SSWithin Group 1 for } XY}{\text{SSWithin Group 1 for } X} + \frac{\text{SSWithin Group 2 for } XY}{\text{SSWithin Group 2 for } X} + \ldots + \frac{\text{SSWithin Group N for } XY}{\text{SSWithin Group N for } X}\right]$$

$$= \sum^{J}\sum^{I}(y_{ij} - \bar{Y})^2 - \left[\frac{\sum^{I}(x_{i1} - \bar{X}_1)(y_{i1} - \bar{Y}_1)}{\sum^{I}(x_{i1} - \bar{X}_1)^2} + \frac{\sum^{I}(x_{i2} - \bar{X}_2)(y_{i2} - \bar{Y}_2)}{\sum^{I}(x_{i2} - \bar{X}_2)^2} + \ldots + \frac{\sum^{I}(x_{ij} - \bar{X}_j)(y_{ij} - \bar{Y}_j)}{\sum^{I}(x_{ij} - \bar{X}_j)^2}\right]$$

$$\Sigma 2 = \text{SSWithin } Y - \frac{(\text{SSWithin } XY)^2}{\text{SSWithin } X}$$

$$= \sum^{I}(y_{ij} - \bar{Y}_j)^2 - \frac{[\sum^{I}(x_{ij} - \bar{X}_j)(y_{ij} - \bar{Y}_j)]^2}{\sum^{I}(x_{ij} - \bar{X}_j)^2}$$

$$\Sigma 3 = \text{SSTotal } Y - \frac{(\text{SSTotal } XY)^2}{\text{SSTotal } X}$$

$$= \sum^{J}\sum^{I}(y_{ij} - \bar{Y})^2 - \frac{\sum^{J}\sum^{I}[(x_{ij} - \bar{X})(y_{ij} - \bar{Y})]^2}{\sum^{J}\sum^{I}(x_{ij} - \bar{X})^2}$$

$$\Sigma 4 = \Sigma 2 - \Sigma 1$$

$$\Sigma 5 = \Sigma 3 - \Sigma 2$$

References

Edwards, A. L. *Experimental Design in Psychological Research (4th ed.).* New York: Holt, Rinehart and Winston, Inc., 1972.

Guilford, J. P., and Fruchter, B. *Fundamental Statistics in Psychology and Education (6th ed.).* New York: McGraw-Hill Book Company, 1978.

Hays, W. L. *Statistics (3rd ed.).* New York: Holt, Rinehart and Winston, 1981.

Kirk, R. E. *Experimental Design: Procedures for the Behavioral Sciences (2nd ed.).* Monterey, California: Brooks/Cole Publishing Company, 1982.

Loftus, G. R., and Loftus, E. F. *Essence of Statistics.* Monterey, California: Brooks/Cole Publishing Company, 1982.

Winer, B. J. *Statistical Principles in Experimental Design (2nd ed.).* McGraw-Hill Book Company, 1971.

Index

239314